Pedestrian
Zones
Car-Free Urban Spaces

Chris van Uffelen

Pedestrian
Zones
Car-Free Urban Spaces

BRAUN

CONTENTS

Take a Walk on the Wild Side

by Chris van Uffelen

Traffic is a city's worst enemy and its best friend. The city needs goods in order to survive, the commuters who work there, the consumers who shop, the tourists who come to gaze in wonder – in short, people who travel in and out of the city. Furthermore, there are also the hundreds of thousands of city residents. Without all of this movement, a city wouldn't be a city. A city needs the hustle and bustle. In the future there is likely to be more people living in cities, so that commuters living on the edge of the city will decrease, but this will simply transfer the problem to the inner city infrastructure, and will thus make little difference. On the other hand, the city as we know it will be characterized by new traffic-free routes and urban open spaces. Almost every large city now has a pedestrian area that forms an important and lively part of the city center. The area around the center and even individual quarters also often have their own pedestrian zones. Car-free areas with top locations for shops and businesses are also popular in small towns and villages.

← | **Souks in Marrakesch: historical streets that cannot be accessed by car or bike**

Most pedestrian zones with businesses are largely car-free during the day, although deliveries are permitted in the early morning. There is even a very early example of this: In 45 BC Julius Caesar declared the Lex Iulia municipalis in Rome to be a wagon-free area during the day. Until the third century AD, no wagons were permitted to enter the city center between dawn and the tenth hour of the day (around 5pm): The only exception to this rule were carnivals and priests, wagons with construction material and waste disposal. The wagons were permitted to enter the city center at night and made an incredible noise driving over the cobblestones.

During the Middle Ages and the Early Modern era, there were often numerous carriages in the cities but the problem only became as acute as it was in the ancient metropolis after mass motorization following the invention of the automobile. Since the 1920s/1930s cars have played an increasingly important role in the planning of future cities. In 1928 Clarence Stein and Henry Wright designed the first car-friendly residential estate with a separate pedestrian entrance in Radburn, New Jersey: The

suburb was born. With his design for Broadacre City in the 1930s–1950s Frank Lloyd Wright planned a large-scale suburb. Later (Athens Charter), city planners began to dissolve traditional city streets with their mix of functions in every city quarter, this in turn increased the volume of traffic. Living outside the city became fashionable in the 1960s/1970s and that generation was the first to find itself trapped not only by architectural developments on one side of the sidewalk, but also a metal wall of cars on the other. It was at this time that people began to view the ever-increasing number of cars more skeptically: commuters spent hours stuck in traffic, artists encased cars in concrete, and cities began building the first pedestrian zones where it was possible to go shopping or simply for a stroll. 1953 was a turning point for these new introductions to the cityscape: Rotterdam opened the Lijnbaan in October, designed by Jo van den Broek and Jacob B. Bakema. After that things began to develop rapidly in Germany: In November, the Treppenstraße was opened in Kassel, followed in December by Holstenstraße in Kiel, and then Schulstraße in Stuttgart. In the 1950s Morris Lapidus

designed Lincoln Road in Miami Beach, the first US American pedestrian zone, which he ornamented with various small architectural elements – as was also common in park design. In the 1960s Jan Gehl embarked upon a pedestrian-friendly renovation of Copenhagen and is even today often consulted worldwide and also designs projects focusing on how to transform the city into "shared space". In the past there were of course earlier car-free cities (Venice), streets and zones (Mackinac Island, Buenos Aires, Essen), but the intention of giving pedestrians a space suited to their needs, and their speed, first came about in the post war era. Since that time, pedestrian zones have become a magnet of the urban fabric, although over the years they have been adapted to suit a variety of different uses. At the beginning, many pedestrian areas were sealed off, while today interfaces that integrate public transport (trams, subway trains, taxi stands), delivery vehicles (load-

ing times and zones) and motorized individual traffic are being planned. Even pedestrian zones that aren't dedicated to shopping but are nevertheless urban linear routes stretching between two points (in contrast to a square) are developing into their own category.

//

This volume presents current developments within this relatively new form of street design. Large areas that stretch across various city quarters are presented next to compact pedestrian zones in small towns. The latter often pay particular attention to the specific historical context of the location and the particular requirements of the users in order to create the optimal experience for residents and visitors. Many of the projects in this volume are new designs for the first generation of pedestrian zones. Others involve the transformation of traffic-heavy city centers into pedestrian zones that are designed to tempt consumers and raise the

quality of life. Others serve to provide stress-free movement across the city. Within the pages of this book, more than 60 projects provide a varied look at current ambitions.

↖ | **Lesser Ury: Unter den Linden Berlin with a view of Brandenburger Tor,** cars began to characterize the cityscape in the 1920s.
↑ | **Morris Lapidus: Lincoln Road in Miami Beach,** roofs provide shade
↗ | **Jo van den Broek and Jacob B. Bakema: Lijnbaan in Rotterdam,** first pedestrian zone in the world
→ | **Lincoln Road in Miami Beach,** pavilion and landscaping

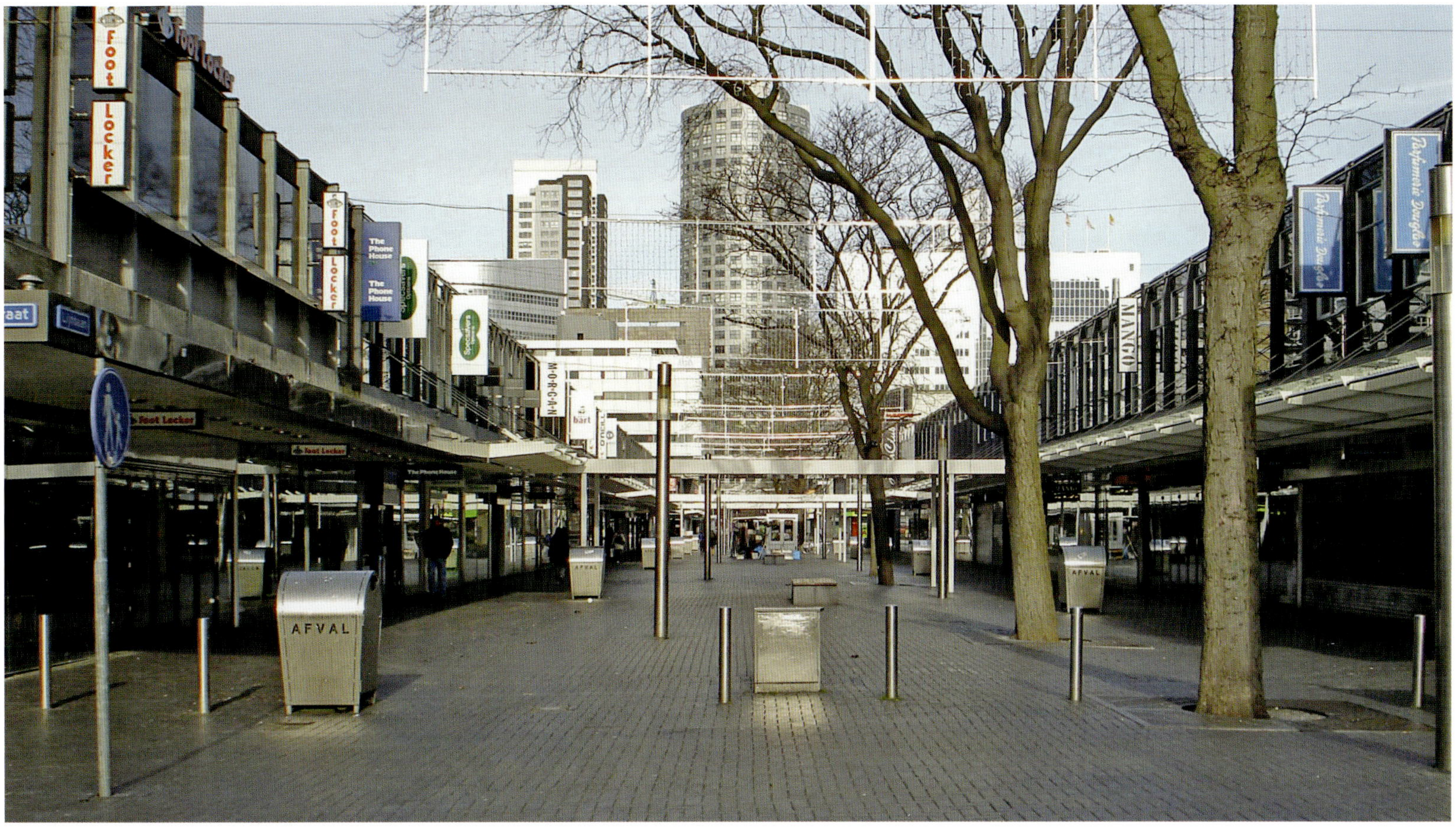
Foot Locker
The Phone House
AFVAL
MANGO

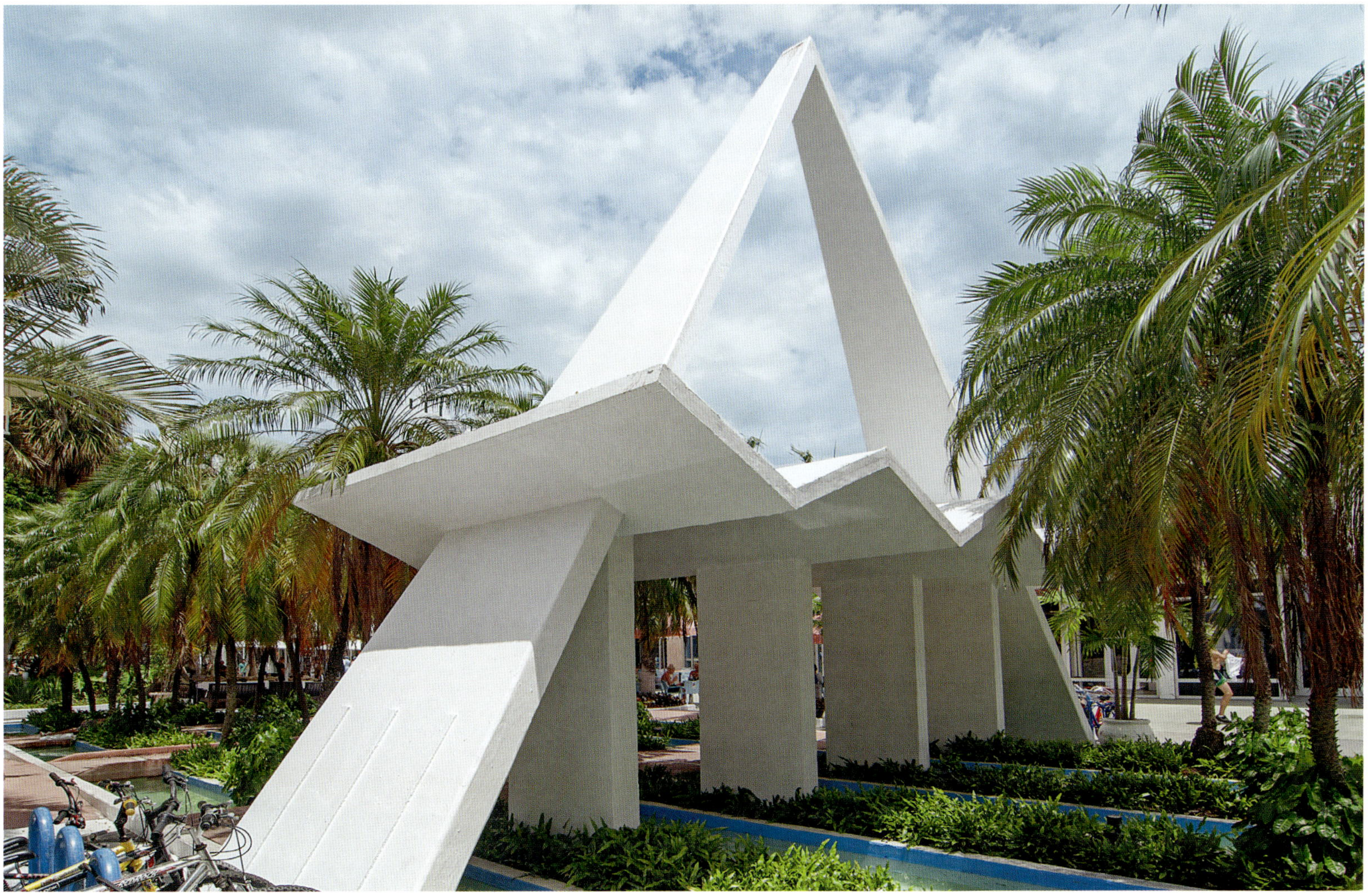

REMODEL

↑ | **View of the Esplanade,** pergola from Atelier Oï
↗ | **La Voie-du-Chariot and Les Mercier's shops,** water feature by Daniel Schläpfer
→ | **Place de Flon-Ville,** tree by Oloom

Quartier du Flon

Lausanne

The Flon district in Lausanne, Switzerland, is a city within a city. Here, art and retail, entertainment and culture, work and enjoyment flow naturally together. In the mid 19th century, the Flon district was home to tanneries and sawmills; during the industrial revolution, the Flon riverbed was filled in and channeled underground. The development of this area has created a 55,000-square-meter, mixed-use paradise on private property in the heart of the city. Its unique topography creates some of the only wide-open, flat pedestrian-friendly spaces in central Lausanne. Improvements continue to be constantly made to the whole district and a redevelopment concept for public spaces is the subject of a future competition. The Flon district is owned and managed by the Mobimo Group.

PROJECT FACTS

Address: Quartier du Flon, Lausanne, Switzerland. **Completion:** 2008. **Artists:** Daniel Schläpfer, Pierre Oulevay, Tina Ausoni. **Length:** 415 m. **Area:** 55,000 m². **Estimated visitors:** 5,000 per day. **Paving:** cobblestone, asphalt, tartan, gore. **Landscaping:** trees. **Street furniture:** design by Oloom and Atelier Oï + 5 Art showcases. **Context:** pedestrian zone in city center.

← | **Place Les Mercier**
↓ | **La Voie-du-Chariot,** diamonds christmas
lanterns by Oloom

↑ | **Three new Pépinières buildings with terrasses**
↙ | **Site plan**

↑ | **Shopping street,** gray granite paving
→ | **Købmagergade shopping street**

Købmagergade Shopping Street

Copenhagen

The curved course of the Købmagergade shopping street is typical of the city center of
Copenhagen. This long street embodies the characteristic image of this labyrinthine medieval city center. The district has its own daily and weekly rhythm: people cycle, walk,
shop, play and go out in the evenings. The first step was to clean and empty the area, so
that the flow of people can be easily channeled. The layout of the three squares is varied,
just as their historical situation and their location in the city are varied. On the Kultorvet
the dark – almost black – paving pattern of the stone is inspired by the 18th-century coal
trade. On the rather more peaceful Hauser Plads square, the exciting grass play mounds
form a green oasis in the urban fabric. At night, the Trinitatis Kirkeplads with its famous
observatory Rundetårn is transformed by artificial lighting into an enormous starry sky
in the floor.

Address: Købmagergade, 1150 Copenhagen, Denmark. **Completion:** 2013. **Area:** 22,000 m². **Paving:** granite in five gray shades. **Street furniture:** by KBP.EU. **Context:** shopping street and three squares in city center.

← | Market
↓ | Pedestrian area at night

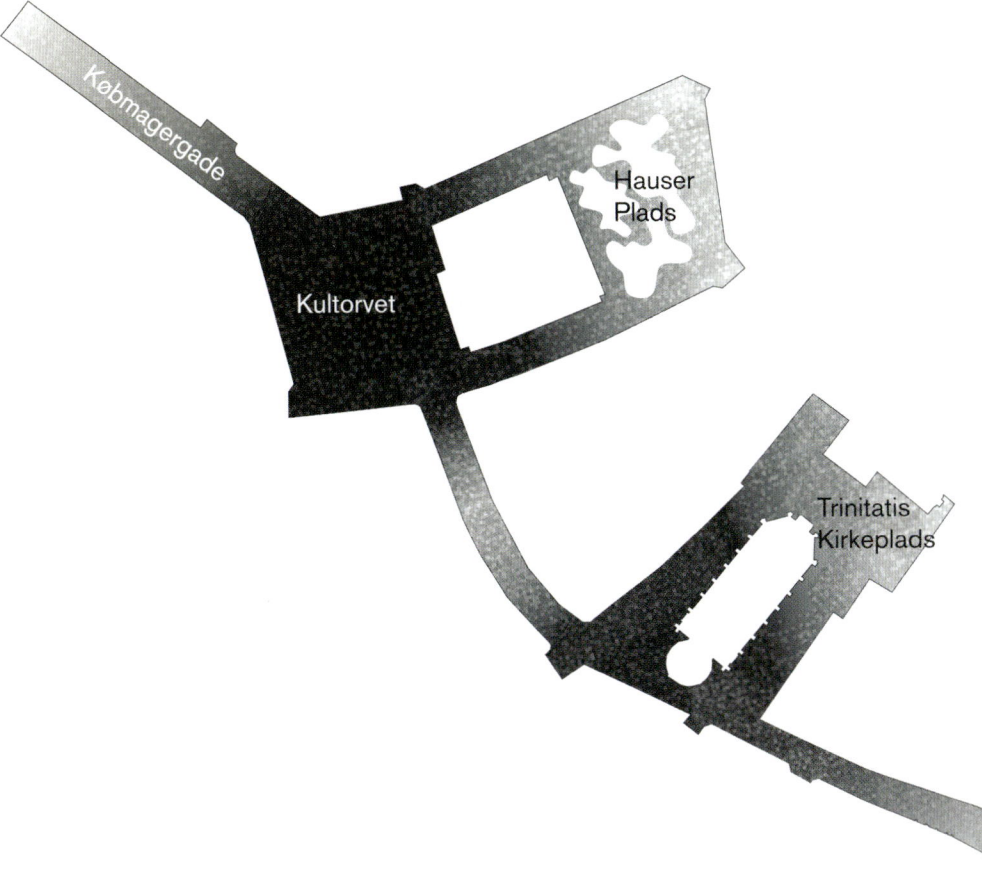

↑ | **Site plan**
← | **Kultorvet Square**, fountain

José Adrião Arquitectos

↑ | **Pink design contrasts historical context**

Pink Street

Lisbon

In December 2011 Nova do Carvalho Street (Rua Nova do Carvalho) in Cais do Sodré was painted pink by local shop owners as an ephemeral installation. The gesture synthesized the need for change. The competition proposal, in September 2012, aimed to strengthen the character established with the previous intervention, giving it continuity and permanency. This design is inclusive, opened and multifunctional, and created a dynamic public space. The proposal consists of leveling the sidewalk to the same height as the road so as to turn the existing space into a barrier-free public space. Along the street, eight billboards are assembled, which can be used for exhibitions or simply to advertise events taking place. The eight billboards can also be used as street lamps.

PROJECT FACTS

Address: Rua Nova do Carvalho, Cais do Sodré, Lisbon, Portugal. **Completion:** 2013. **Length:** 108 m. **Area:** 1,391 m². **Paving:** colored paving. **Context:** Cais do Sodré highstreet renovation.

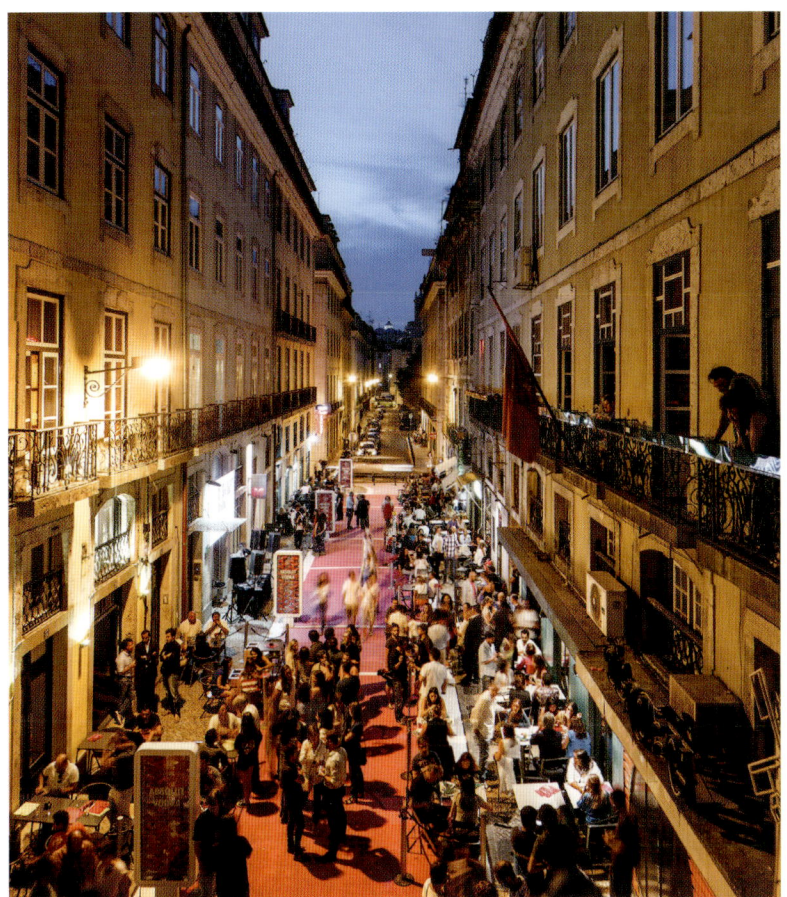

↑ | Bustling street at night

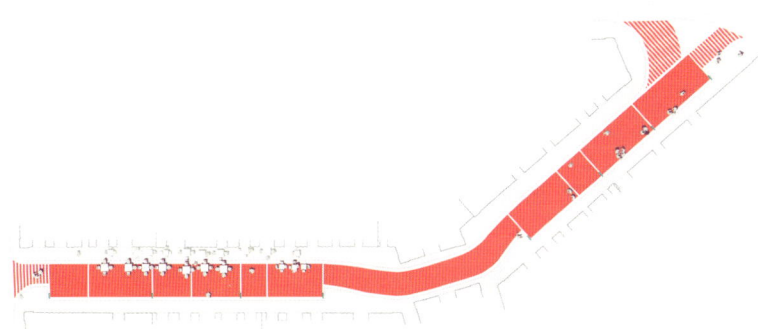

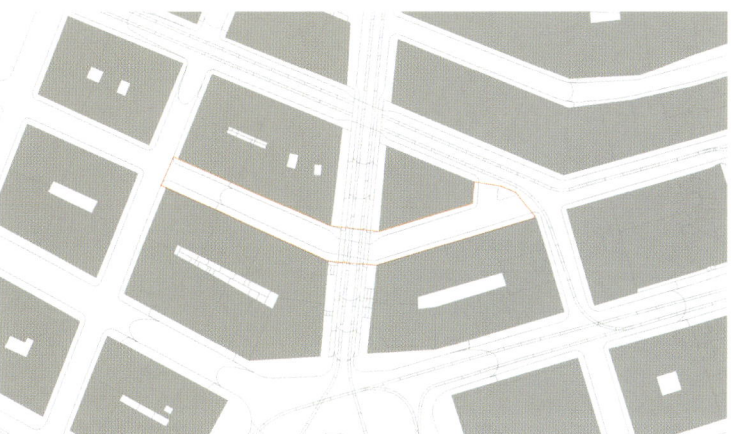

↑ | Layout and site plan

↓ | Colorful combination of white and pink

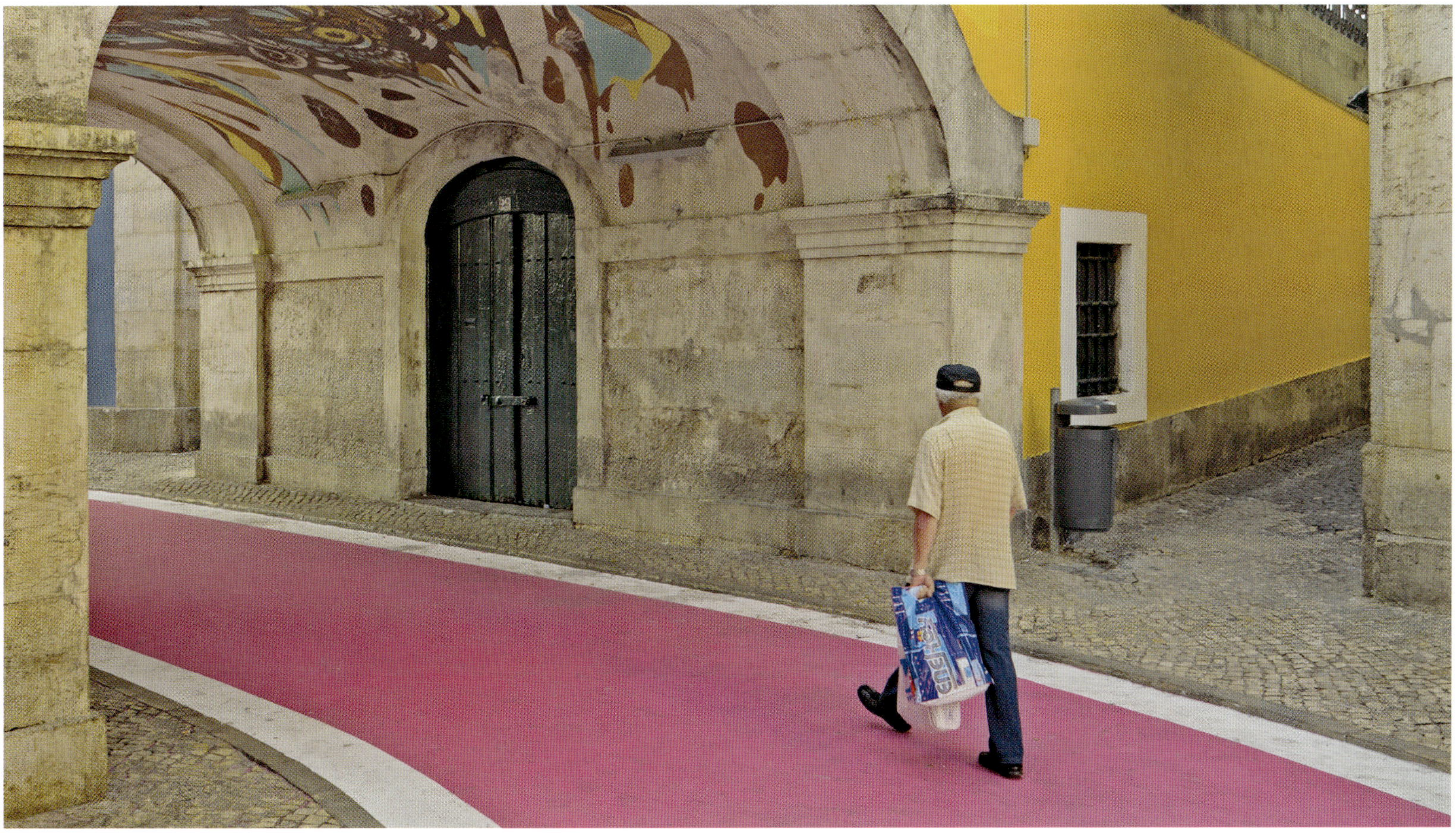

↑ | **Cabot Circus panorama,** semi-open shopping mall

Cabot Circus / Quakers Friars

Bristol

Cabot Circus unites Broadmead, Bristol's original shopping district, and Quakers Friars. Novell Tullett was commissioned to design the public realm for this £300 million, mixed use, city quarter opening in September 2008. The scheme re-established routes to local districts with new pedestrian streets and public spaces. The Quakers' meeting house and former Dominican Friary is at the heart of one space where a light touch was essential to marry heritage with contemporary, understated style incorporating four statuesque existing trees. Locally sourced Welsh pennant stone forms a simple ground plane throughout the scheme with new trees and fountains bringing a sense of vitality. The team's aspiration for a clutter-free highway is achieved in Bond Street – a new arrival space for Bristol.

PROJECT FACTS

Address: Cabot Circus, Bristol BS1 3BX, England. **Completion:** 2008. **Architects:** Chapman Taylor, Alec French Architects, Stanton Williams. **Artist roof of Cabot Circus:** Nayan Kulkarni. **Area:** 33,000 m². **Estimated visitors:** 53,000 per day. **Paving:** Welsh pennant stone and York stone with granite trims. **Landscaping:** more than 140 new fastigiate oak, ash and plane trees. **Street furniture:** Voss Street Furniture. **Context:** shopping street in city center.

↑ | **Philadelphia Street,** pedestrian area uniting Quakers Friars and Cabot Circus

↑ | **New square,** restored Quakers' meeting house
↓ | **Landscape master plan**

Keller Damm Roser
Landschaftsarchitekten
Stadtplaner

↑ | **Exterior,** Sendlinger Straße
→ | **Detail,** planting trough

Hofstatt

Munich

The site, situated in the historic center of Munich, was used by the Süddeutscher Verlag as a publishing house for decades. After the area went out of use, a redesign was conceived for the site. The existing historic buildings, some parts of which are listed, were complemented with new ones to create a unique ensemble, forming a new urban living and business address, the Hofstatt, in the center of Munich. The design concept for the new buildings constitutes a clear distinction between the different time layers. On the ground floor, a three-branch retail passage forms the core of the building volume. The connected courtyards provide open space, each with a character of its own. The informal mix of public areas and restaurants creates a relaxing place to spend time.

PROJECT FACTS

Address: Sendlinger Straße 8–12, 80331 Munich, Germany. **Completion:** 2013. **Architects:** Meili, Peter GmbH, Munich. **Landscape architects:** Vogt Landschaftsarchitekten, Zurich (competition), Keller Damm Roser (realisation). **Length:** 200 m. **Area:** 2,000 m². **Paving:** different paved surfaces of natural stone. **Context:** shopping center.

← | **Private yard**
↓ | **Site plan**

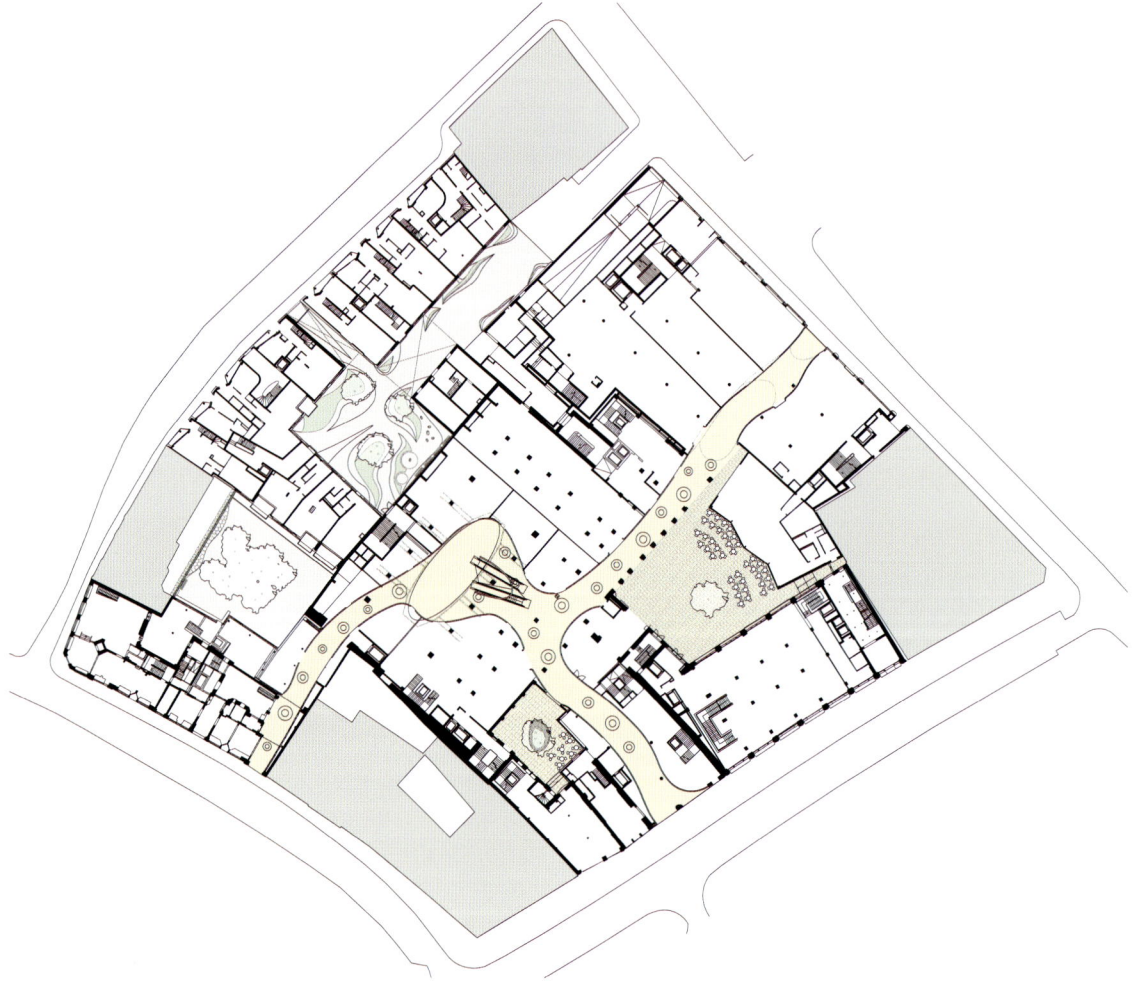

↑ | Courtyard
← | Trough for plants

Behnisch Architekten

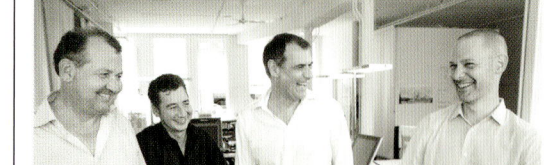

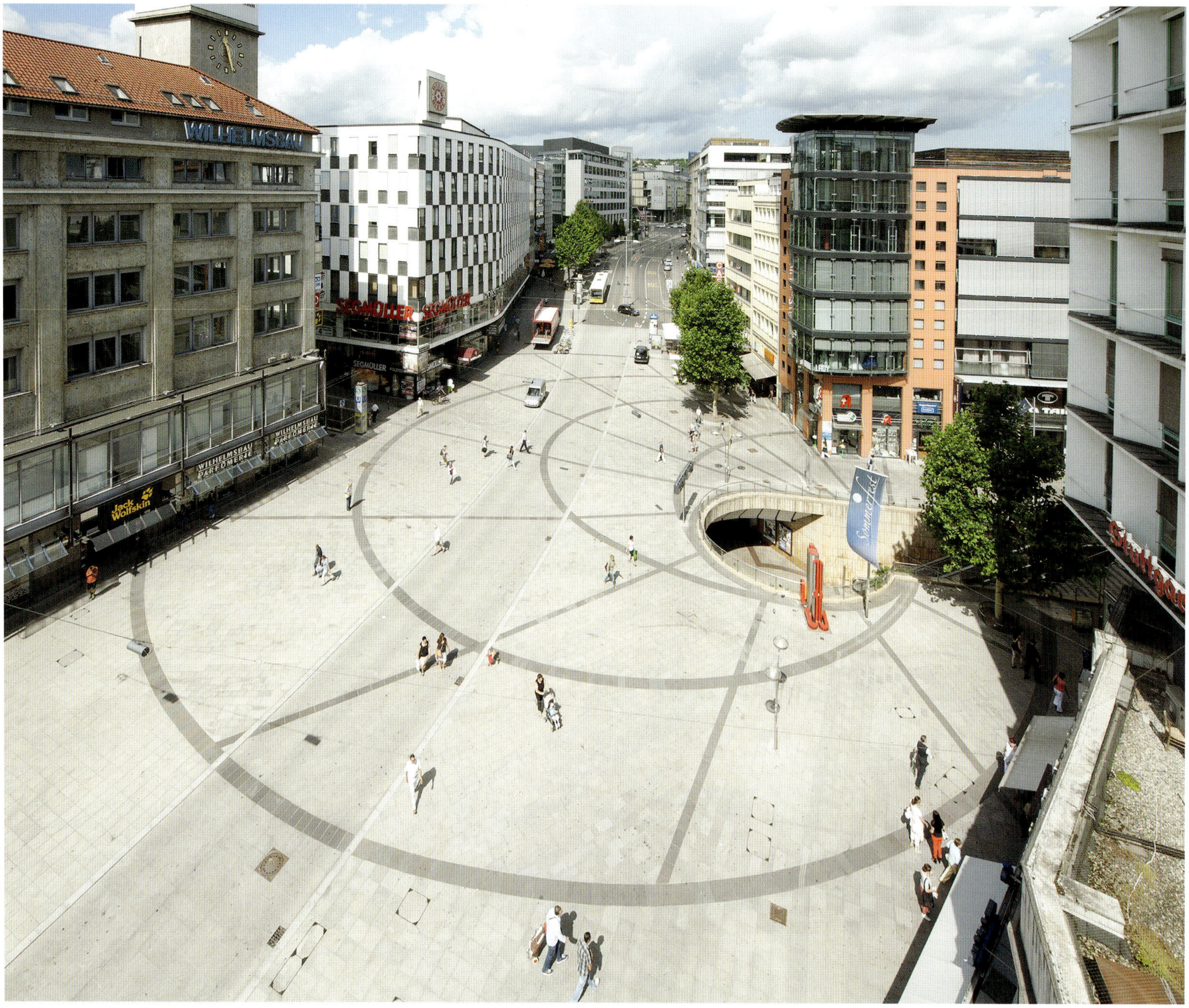

↑ | Link road transformed into pedestrian area
→ | Bird's-eye view

Redesign of Königstraße

Stuttgart

In the 1970s Behnisch & Partner transformed the Königstraße into a pedestrian zone. Together with the neighboring Schlossplatz, this forms the central element of the public space in the heart of the city. After 30 years the area was in need of renovation and modernization work, which was carried out in several phases between 2005 and 2011. The entire length of the Königstraße has been paved with a unified surface of Flossenbürger granite paving stones. Towards the center bands of light colored shell limestone can be easily differentiated from the granite. In the second to last building phase the link road at the south end of the Königstraße was transformed into a pedestrian area. This is intended as an attractive address for retailers and is now better integrated into the inner city.

PROJECT FACTS

Address: Königstraße, 70173 Stuttgart, Germany. **Completion:** 2011. **Length:** 1,200 m. **Area:** 40,000 m². **Estimated visitors:** 100,000–150,000 per day. **Paving:** Flossenbürger granite, shell limestone bands. **Landscaping:** plane trees. **Street furniture:** integrated seating, play and water elements. **Context:** shopping street in city center.

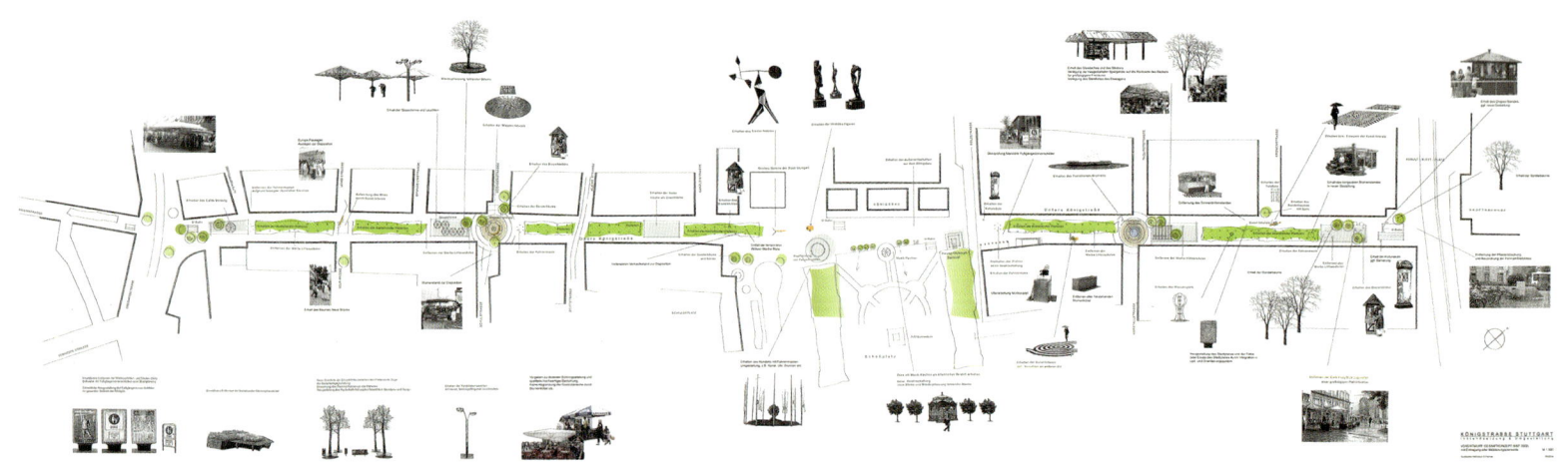

↑ | **Master plan**
↓ | **Plane trees line main shopping street**

↑ | **Königstraße with protective umbrellas and play elements**
← | **Site plan**

birke . zimmermann
landschaftsarchitekten

↑ | Mixture of trees, flowers and grasses
↓ | Weasel play elements

↙ | Water features
↓ | Bench detail
↗ | Pedestrian zone lined with shops
→ | Landscaping detail

Pedestrian Zone

Wesel

The Hanseatic city of Wesel was so badly damaged during World War II that historical architecture simply no longer exists in the inner city. The main goal of this project was to create a modest yet characteristic design, which would offer plenty of space for interaction and unify the varying architectural styles and city spaces. Large paving slabs and tiles draw the various areas together and form a 700-meter-long pedestrian zone that connects the two most important areas of the city – Großer Markt and Berliner Tor Platz.

PROJECT FACTS

Address: Hohe Straße, 46483 Wesel, Germany. **Completion:** 2013. **Area:** 18,000 m². **Paving:** concrete paving stone with natural stone. **Landscaping:** Raywood ash mixed with various herbaceous perennials. **Street furniture:** benches, rotating benches, play elements. **Context:** shopping street in town center.

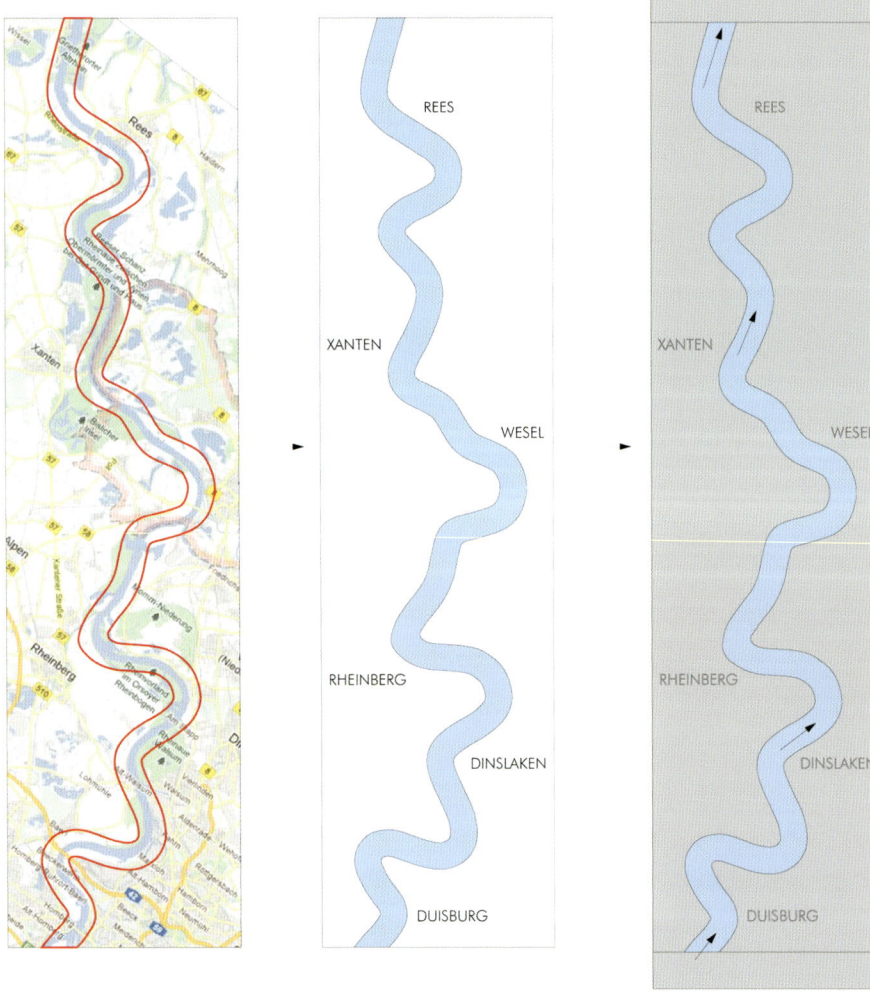

← | **Water feature,** concept plan
↓ | **Water feature**

↑ | **Wooden benches**
← | **Sculptural water feature,** designed to look like a river

↑ | **Buchanan Street,** evening
→ | **Landscaping brightens up the urban space**

Buchanan Street

Glasgow

Following Gillespies' adopted public realm strategy for Glasgow's city center, the practice won an international design competition for Buchanan Street – one of Glasgow's central and most famous streets. This project involved a reduction in vehicular access and the creation of a pedestrian-oriented environment. The practice produced a design for the street based upon the rich embedded vernacular traditions of the city, interpreted in a modern, clean and functional manner. As well as using traditional, robust and durable natural stone paving materials, simple and elegant elements such as lighting and seating components were developed to create a strong rhythm along the street. The project has won recognition for its quality both in the United Kingdom and internationally, winning numerous awards.

PROJECT FACTS

Address: Buchanan Street, Glasgow, Scotland. **Completion:** 2006. **Area:** 10,000 m². **Paving:** granite slabs. **Landscaping:** semi-mature street trees. **Context:** shopping street in city center.

↖ | Sketch
↙ | Steps provide a sunny place to sit
↓ | Lighting design gives street a more modern appearance

↑ | Trees line the street
← | Looking down Buchanan Street

BPG Landschaftsarchitekten

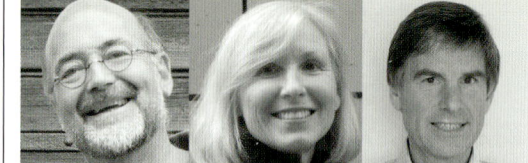

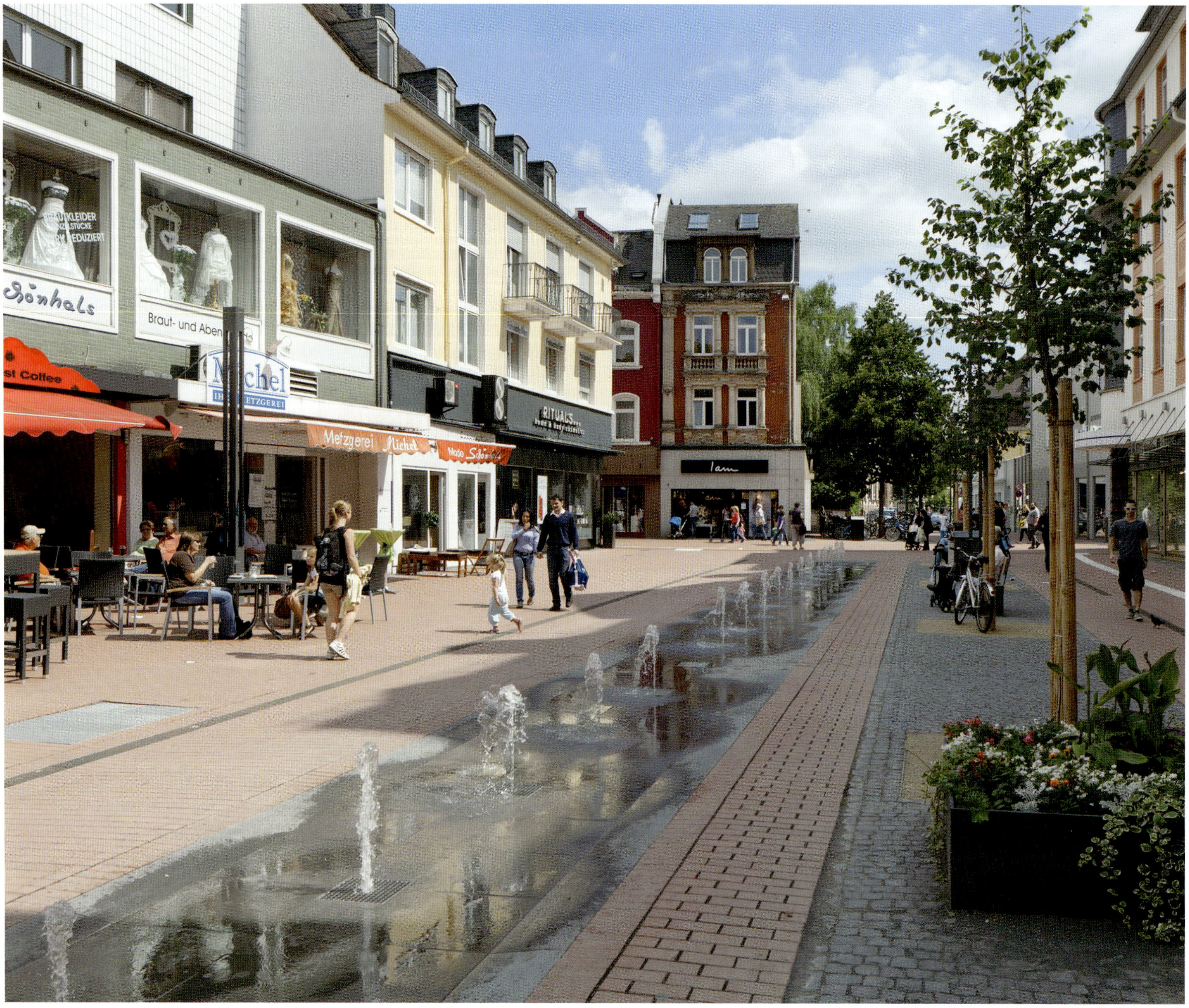

↑ | **Löwengasse,** water feature
→ | **Seltersweg**

Pedestrian Zone

Gießen

This project is the winning entry to a competition to redesign Gießen city center. The concept rejuvenates the street space by the integration of longitudinal patterns in the street surface. This responds to the classic organization principle of a street and became one of the leading motifs for the design. The alternation of greenery and open areas, combined with high-quality paving, gives the inner city a new character and transforms it into a welcoming place that encourages both residents and visitors alike to linger. The design of the crossroads and public squares are particularly eye-catching. Fountains, seating elements and individual play elements come together to create an attractive inner city space.

PROJECT FACTS

Address: Seltersweg, 35390 Gießen, Germany. **Completion:** 2014. **Sculptural fountain:** Ruth Leibnitz. **Length:** 900 m. **Area:** 16,000 m². **Estimated visitors:** 18,000 per day. **Paving:** clinker, basalt, concrete. **Landscaping:** evergreen oak trees, hornbeams, Turkish hazel. **Street furniture:** seating elements, flower beds, tree pits, individual play elements. **Context:** shopping street in city center.

← | **Structural pavement sections,** row of benches, trees and flower beds
↓ | **Seltersweg at night**

↑ | Site plan
← | Fountain with globes

BHF
Bendfeldt Herrmann Franke
Landschaftsarchitekten

↑ | **Main pedestrian route**
↗ | **Pedestrian route**, links town and coast
→ | **Waterfront**

Königstraße and Sandwall Pedestrian Zone

Wyk auf Föhr

Established in 1819, this is the oldest seaside resort on the North Sea coast in Schleswig-Holstein. The Sandwall with its double row of trees and the Königstraße are the central elements connecting the bastion to the west with the Rathausplatz to the east. The pedestrian zone is located directly on the North Sea coast and leads from the Königstraße to the Sandwall. The Königstraße is named after the Danish king Christian VIII, who used the area as his summer residence from 1842–1847. The old layout from the 1970s has been replaced with a band of concrete plaster accompanied by a gray band of granite paving. The design by BHF Bendfeldt Herrmann Franke Landschaftsarchitekten creates a welcoming location that emphasizes the town's connection to the sea and beach.

Address: Königstraße and Sandwall, 25938 Wyk auf Föhr, Germany. **Completion:** 2009. **Length:** 600 m. **Area:** 10,000 m². **Paving:** concrete, granite. **Context:** shopping street in coastal town center.

← | Bustling shopping street
↓ | Site plan

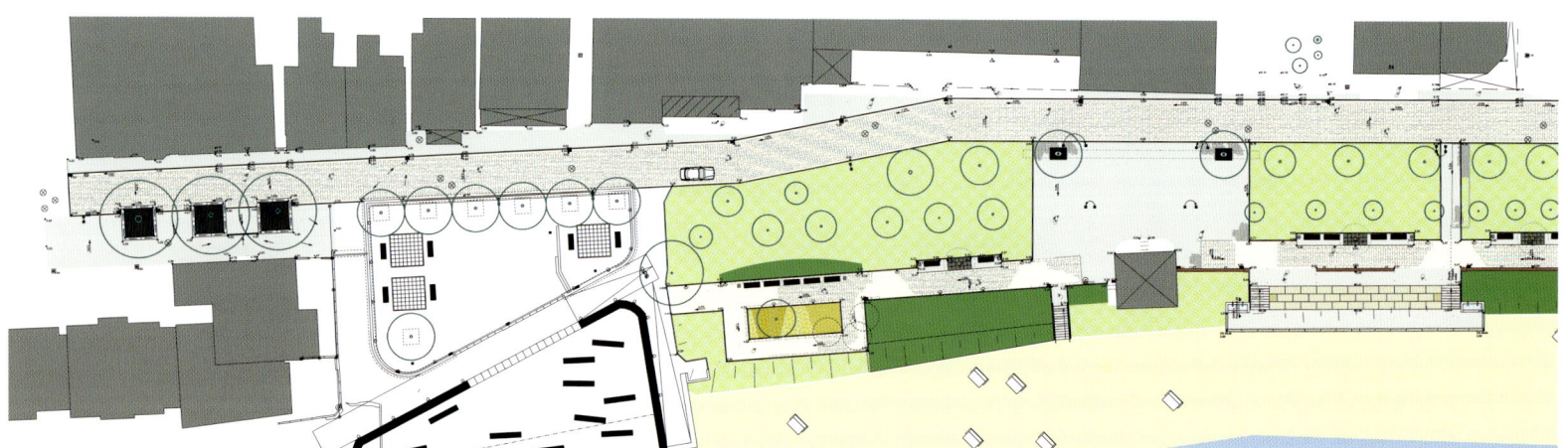

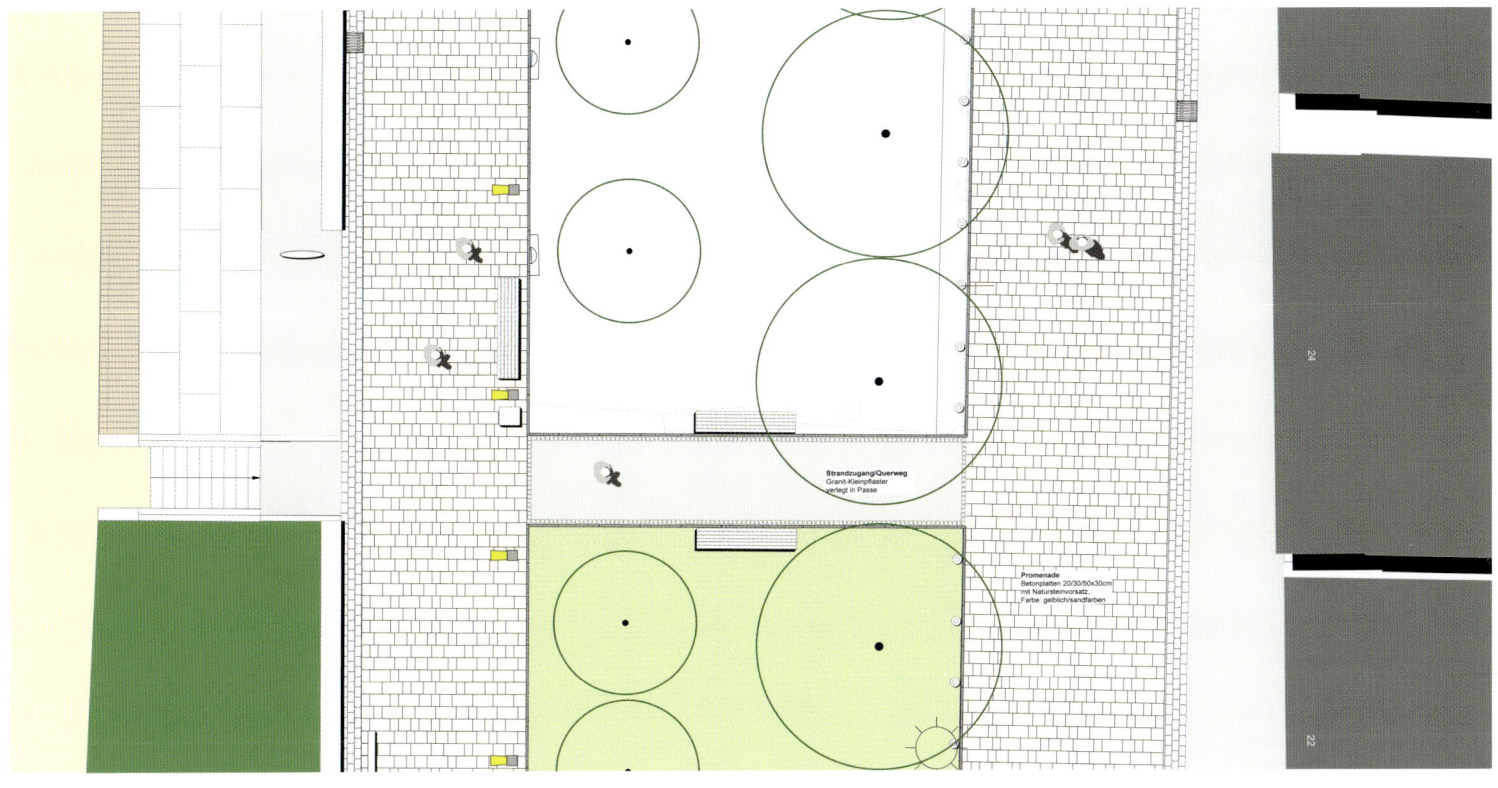

↑ | **Site plan**, detail
← | **New layout makes the most of the coastal views**

Isthmus Group

↑ | **Artistic street furniture,** colorful accent in streetscape

Osborne Street

Auckland

Isthmus was commissioned by Auckland City Council to provide a strategic development plan for Newmarket, Auckland's main retail and fashion center. The plan developed a hierarchy for the streets within the wider central business district and included a Public Realm Design Guide with specific treatments for curbs, paving, vegetation and street furniture. Previously a run-down side alley, Osborne Street is one of a number of smaller scale lanes off Broadway, which have been redeveloped as high quality streets with boutique retail outlets inhabiting restored industrial buildings. Osborne Street included full kerb and carriageway redesign and rationalized parking that provides a naturally regulated shared space. The integration of public art was at the forefront of the design objective of the wider project, contributing to the character of the individual spaces.

PROJECT FACTS **Address:** Osborne Street, Newmarket, New Zealand. **Completion:** 2010. **Length:** 184 m. **Area:** 2,539 m². **Paving:** Asian basalt and brown granite. **Artist:** Seung Yul Oh. **Context:** redevelopment of main retail and fashion center.

↑ | **Street lined with shops**

↑ | **Sketch,** connection between Broadway and Osborne Street
↓ | **Basalt paving references underlying volcanic bedrock**

WES LandschaftsArchitektur
Hans-Hermann Krafft

↑ | **Colorful historical buildings,** complemented by paving
→ | **Old town hall**

Weender Straße Pedestrian Zone

Göttingen

Weender Straße is Göttingen's most important town center axis, characterized by its elegance and timeless design. Visitors are greeted by a homogenous 'carpet' of large natural paving stones in light beige and red-brown tones that reaches across the entire street. The colors present on the historical buildings appear to be reflected by the subtle changes in the paving. The appearance of the street changes depending on the time of day and the weather. A key element of the design is the inclusion of the old town hall into the design of the public area. A 'carpet' of paving stones the same size as the town hall's floor plan structures the area and emphasizes the presences of the town hall.

PROJECT FACTS

Address: Weender Straße, 37073 Göttingen, Germany. **Completion:** 2014. **Length:** 300 m. **Area:** 13,500 m². **Estimated visitors:** 5,000 per hour. **Paving:** granite. **Landscaping:** single trees. **Street furniture:** benches and lighting. **Context:** shopping street in town center.

←← | **Beige and red-brown tones respond to architecture**
↑ | **Street divided by wide drains and borders**
← | **Site plan,** detail

In Situ Atelier de paysage et urbanisme

↑ | Pedestrian zone in historical town center

Pedestrian Area

Saint-Flour

Three streets and four squares have been built on different scales and serve different functions. A simple but sensible project that spreads throughout the city center, these public spaces offer a continuous vocabulary of varying shapes and materials. The identity of Saint-Flour lies in its medieval uptown. In the meantime, an urban renewal project has been commissioned to study the status of the downtown area as a gate to this historic center. The 'Place de la Liberté' urban project is part of a strategy to transform this former 'Faux-bourg', or borough, into a real quarter of Saint-Flour.

PROJECT FACTS

Address: 15100 Saint-Flour, France. **Completion:** 2006. **Lighting designers:** LEA. **Length:** 2,000 m.
Area: 16,500 m². **Estimated visitors:** 130 per day. **Context:** historical city center.

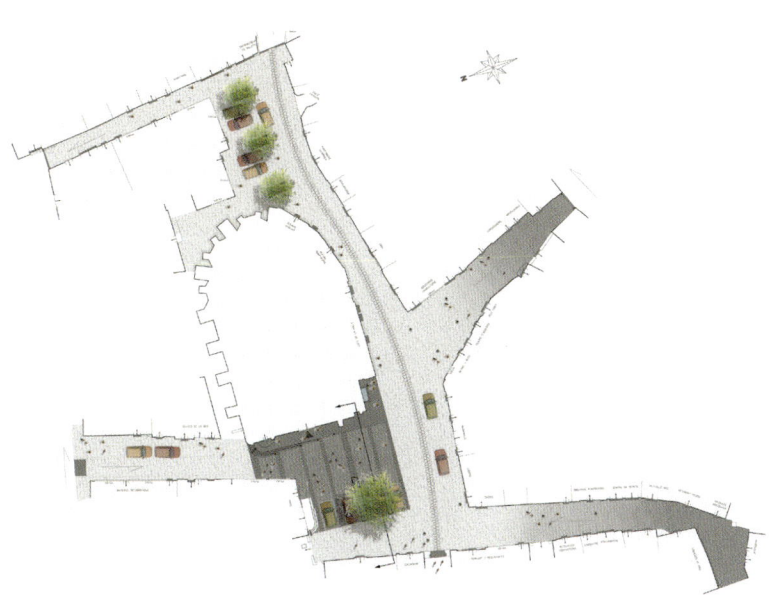

↑ | **Site plan**

↑ | **Steps,** paving detail
↓ | **Pedestrian path winds past historical castle**

Martínez Lapeña Torres
Arquitectos

↑ | **Square with columns**
→ | **Two Roman columns**

Alameda de Hércules

Seville

In 1574, the Laguna de la Feria became a public space, and when two Roman columns, with Hercules and Julius Caesar standing on their capitals, were placed at its southern end: It became known as Alameda de Hércules. In 1765 two new columns supporting two lions with the coats of arms of Spain and Seville were placed at its northern end. The area for vehicular circulation is limited to give priority to pedestrians and to the terraces for the bars. New gentle slopes improve the views of open-air performances. Two kiosks are accompanied by other similar ones with triangular plans. Water spray escapes from the joints between the enamelled paving of three stains in blue and white. The "flamenco and bull-fighter corner" gathers the statues of well-loved personalities from the neighborhood.

PROJECT FACTS **Address:** Alameda de Hércules, Seville, Spain. **Completion:** 2008. **Area:** 37,707 m². **Paving:** sand-colored bricks. **Context:** historical pedestrian zone.

← | Pavilion with seating
↓ | Boulevard lined with trees

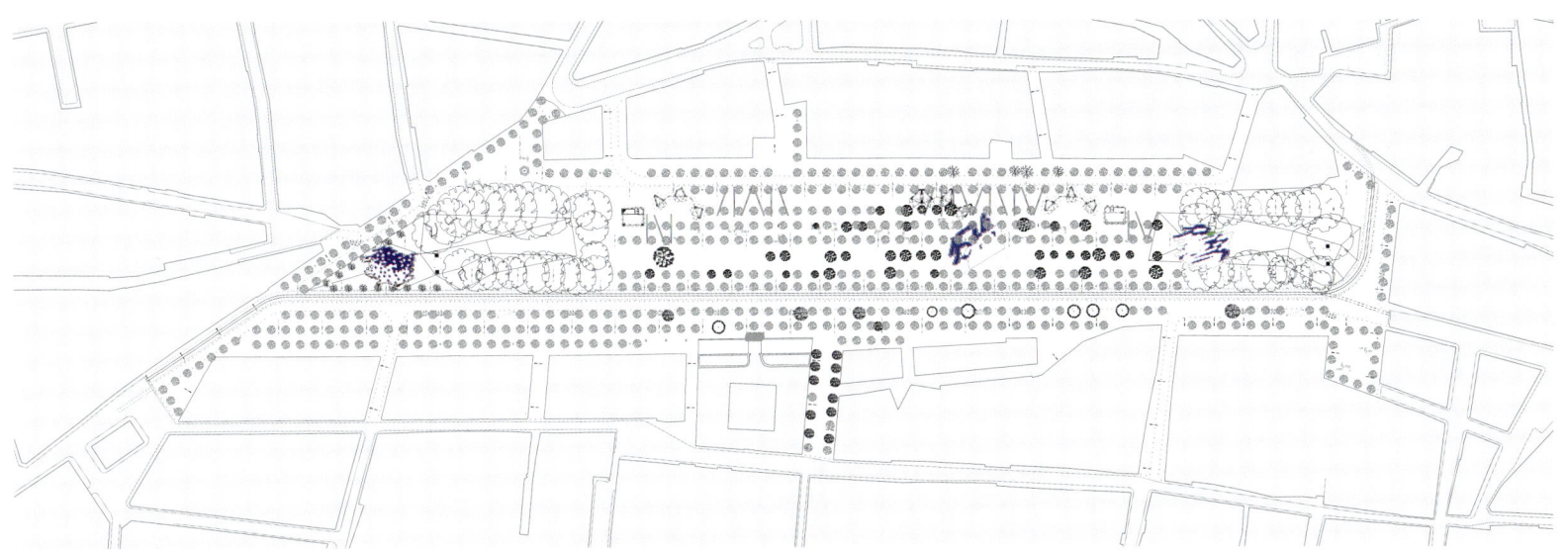

↑ | **Wide car-free boulevard,** a safe place to walk and play
↓ | **Site plan**

r+b landschafts architektur

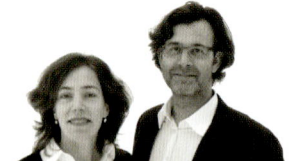

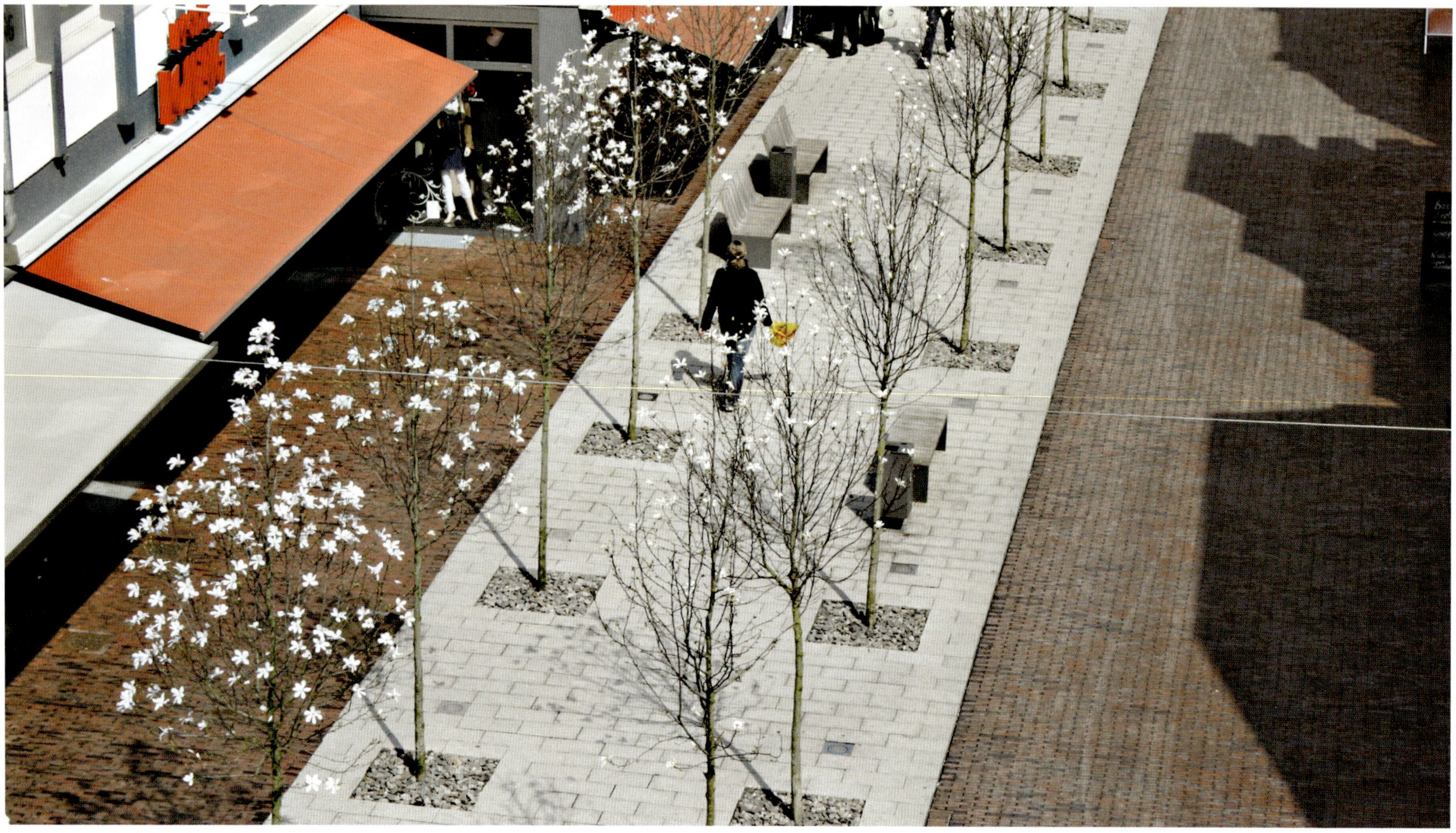

↑ | **Magnolia grove,** bird's-eye view
↓ | **Detail,** water feature

↙ | **Pedestrian zone at night**
↓ | **Detail,** seating accommodation around tree

Pedestrian Zone Modernization

Lingen

The renewal of this town center was not intended to be an historicizing project, but rather a timeless impulse to promote optimization and breath new life into the urban center. The care and careful carving out of regionally typical images was a main focus of the design, with particular emphasis placed on cleaning up, removing and reorganizing some of the original elements. The color and material concepts reference the colors that characterize the town, the red and white of the brick façades and the wood tones of the timber framed buildings. The newly designed surfaces, combined with large bands of granite paving, are complemented by the street furniture. The characteristic design integrates the various structural developments seen in the area.

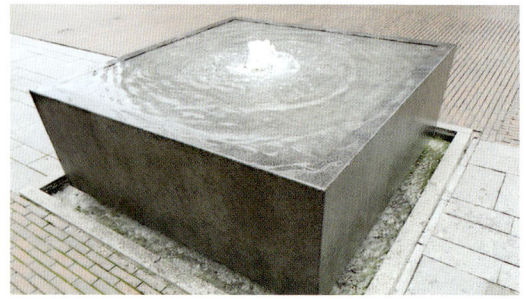

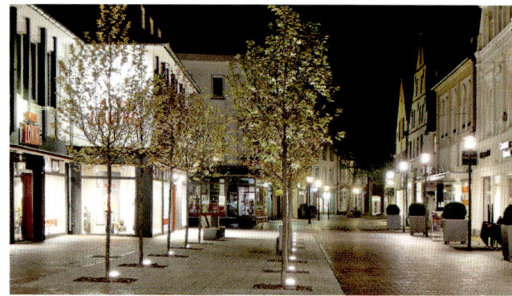

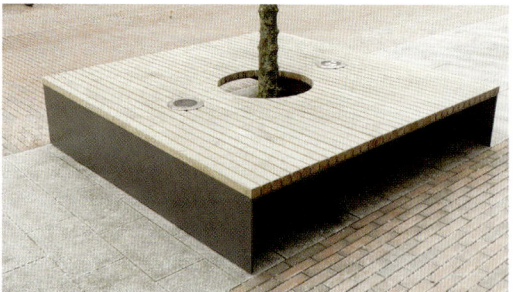

Address: 49808 Lingen (Ems), Germany. **Completion:** 2014. **Length:** 1,310 m. **Area:** 21,385 m². **Estimated visitors:** 16,000 per day. **Paving:** clinker, granit. **Landscaping:** Magnolias, Liquidambar, single maple trees, Amelanchier and Buxus in planters. **Street furniture:** drinking fountain, water features, child's play, benches, trash cans. **Context:** pedestrian area in historic city center.

↑ | Site plan

↑ | **Magnolia trees,** in full bloom
↓ | **Street furniture,** trees provide shade

↑ | **Sunny boulevard,** green landscaping
→ | **Benches and street furniture,** optimally positioned near the landscaping

Remodeling of Passeig de Sant Joan Boulevard

Barcelona

This new remodeling proposal sets two basic objectives: to give priority to the pedestrian use of the boulevard and to turn it into a new urban green zone. In order to achieve these objectives the project has adopted three fundamental urban planning criteria; aiming to guarantee the continuity of the section along the length of the boulevard, to adapt the urban spaces to different uses, and to promote the area as a new and sustainable urban green zone. The new proposal gives the Passeig de St Joan back its social identity as an urban space. This street's urban transformation has enabled the landscape architects to revitalize its commercial life and recreational uses, whilst at the same time retrieving its historical value as a main boulevard that continues right up to Ciutadella Park.

PROJECT FACTS
Address: Passeig de Sant Joan, 08010 Barcelona, Spain. **Completion:** 2011. **Area:** 31,455 m². **Context:** green zone in urban context.

← | Playground
↓ | Street design in context

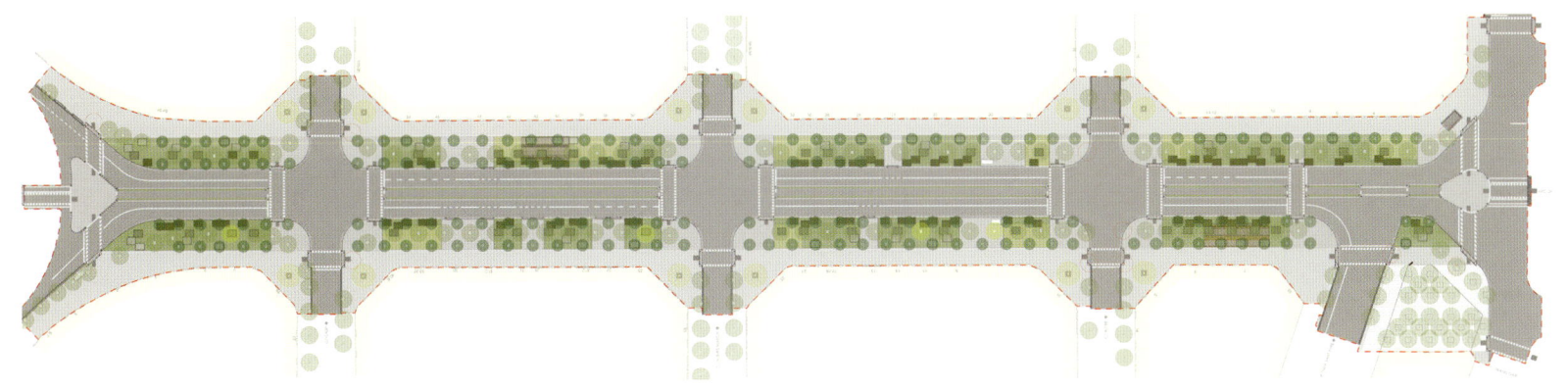

↑ | Master plan
← | Lush green oasis in the middle of the
bustling city

↑ | Bustling shopping street
→ | Cyclist friendly design

New Road

Brighton

Until 2008, New Road was a run-down, under-used street dominated by traffic and little used by pedestrians, despite its location at the heart of Brighton. Landscape Projects, working with Gehl Arkitecten were commissioned by Brighton and Hove City Council to transform New Road into a pedestrian friendly place; the focus of Brighton's Cultural Quarter. The design of the street includes a continuous level surface of varying tones of granite, with white granite edgings and channels defining site features. Surface materials and furniture are arranged to promote pedestrian activities and street clutter has been reduced to a minimum. The New Road project has transformed this part of Brighton, making a socially and environmentally sustainable public realm place.

PROJECT FACTS

Address: New Road, Brighton, England. **Completion:** 2008. **Artist:** Ester Rollinson. **Length:** 170 m. **Area:** 2,800 m². **Paving:** granite. **Street furniture:** wooden benches. **Context:** transformation of inner-city pedestrian zone.

← | **Wide street,** shops and cafés
↓ | **Site plan**

↑ | Integrated seating
↙ | Visualization of scheme

↑ | **Atmospheric blue lights at night**
→ | **Bird's-eye view,** the Corso

Manly Corso

Sydney

Walking down Manly Corso ends one of the world's great journeys; to travel from Circular Quay in the Sydney City, across the harbor by ferry to Manly Beach. While being a highly visited tourist destination, Manly Corso is a critically important part of the character and ambience of the Manly Village, holding special significance to Manly locals for cultural, economic and social reasons. Manly Corso represents a microcosm of Manly life, and the design seeks to embody those characteristics: friendly, egalitarian, laid back. As such, new elements are introduced to the Corso to provide a strong, boldly designed urban environment that allows the space to be read and understood easily, to assist way-finding for the first time visitor, but also meet the expectations of the Manly locals by continuing to provide them with a civic and cultural space they can still call their own.

PROJECT FACTS

Address: The Corso, Sydney 2095, Australia. **Completion:** 2008. **Length:** 200 m. **Area:** 7,800 m². **Estimated visitors:** 1,346 per day. **Paving:** granite. **Landscaping:** palm trees. **Street furniture:** designed by TCL and Dryden Crute Design. **Context:** city corso in touristic context.

← | **Custom-designed curved stainless steel seats,** echo the paving pattern
↓ | **Pavement design,** waves and palms

↑ | **Site plan**
← | **Corso at beach end,** combination of different lights

Jörn Wagner
Freier Landschaftsarchitekt

↑ | **View from townhall square to Zingelstraße**
↘ | **Site plan**

↓ | **Brooch of Meldorf**

Pedestrian Zone Redevelopment

Meldorf

The new design of this pedestrian zone is intended to improve the historical structures as well as creating an unique atmosphere. The street space is kept largely free of fixtures and unified by a continuous paving surfaces. The area has been developed into a welcoming street that invites residents and visitors alike to enjoy an afternoon of window-shopping. The street junctions are emphasized by integrated paving slabs, which depict the "Brooch of Meldorf", historical proof of an early human settlement. The lighting concept pays particular attention to creating a special atmosphere at night.

PROJECT FACTS

Address: Spreetstraße, Roggenstraße, Zingelstraße, 25704 Meldorf, Germany. **Completion:** 2013.
Construction manager: Bornholdt Ingenieure. **Length:** 500 m. **Area:** 4,415 m². **Paving:** concrete,
granite. **Landscaping:** Carpinus betulus (hornbeam), Sorbus intermedia (Swedish whitebeam). **Street
furniture:** furniture series "Plaza" and street lights "Marit" designed by Jörn Wagner, art objects at town-
house plaza created by Johannes Michler. **Context:** pedestrian zone in city center.

↑ | **Granite seating elements,** integrated lights

↑ | **Town hall square**
↓ | **Pedestrian zone Meldorf,** town hall square

hirner & riehl architekten
und stadtplaner

↑ | **Atmospheric evening lights**
↗ | **Alley of plane trees**
→ | **Playground,** water feature

Remodeling of Pedestrian Zone

Bayreuth

1,000 umbrellas for Bayreuth – that was the theme of this prize-winning design to rede-sign Bayreuth's inner city. The basis for the design was a desire to improve the quality of the space and transform it into an area where both visitors and residents would want to spend time. Green, water, and light are the three fundamental design elements that charac-terize the transformation of the pedestrian area. Additionally, the Bayreuth fried sausage stand has been given a new modern interpretation, based on the idea of a classic house form. Together with the rejuvenated urban center, the inner city is now an attractive, vital and welcoming city space.

PROJECT FACTS

Address: Maximilianstraße, 95444 Bayreuth, Germany. **Completion:** 2011. **Lighting designers:** day & light Lichtplanung. **Length:** 550 m. **Area:** 12,500 m². **Paving:** granite. **Landscaping:** Gleditsia, alley of plane trees, flower boxes. **Street furniture:** unitary sunshades, sausage stand, info booth, street furniture by hirner & riehl. **Context:** pedestrian zone remodeling in historical city center.

← | **Bird's-eye view of Maximilianstraße**
↓ | **Site plan**, pedestrian zone

↑ | View towards market
← | Seating island

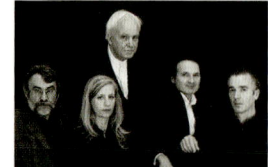

WES LandschaftsArchitektur

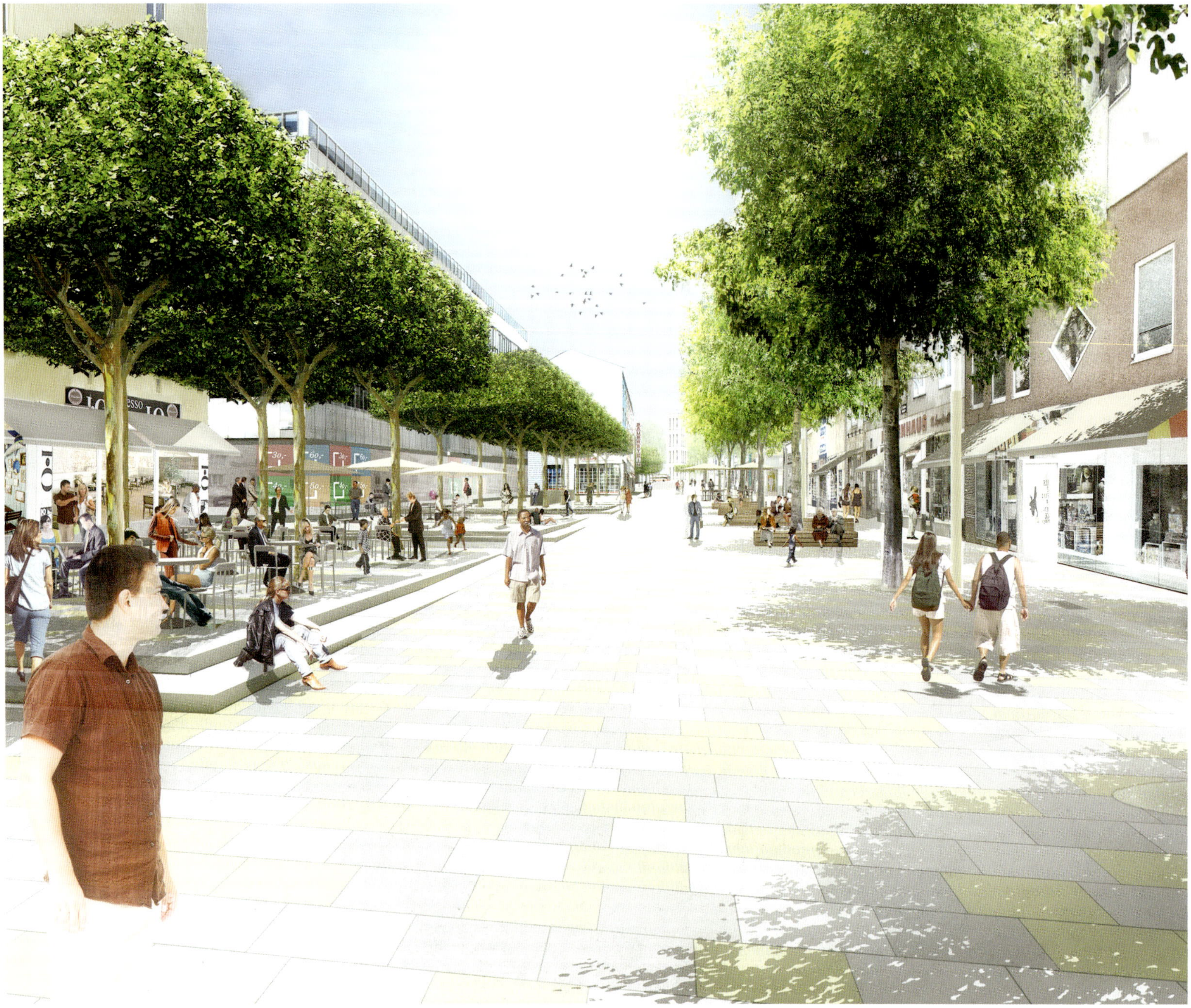

↑ | **Meidlinger Hauptstraße,** plane tree grove
→ | **Meidlinger Platzl**

Meidlinger Hauptstraße

Vienna

The Meidlinger Hauptstraße is Vienna's fifth largest shopping street. This renovation project has been developed as an urban design that responds specifically to the place. The high-quality space invites passers-by to sit and relax, while maintaining a highly functional and useable character. A new feature of the Meidlinger Hauptstraße is its dual function as open public space for residents and as an attractive shopping street. The central design element is a unified, light carpet of ocher, white, and gray granite paving slabs. This creates a stylistic world that is bathed in a yellow-toned light.

PROJECT FACTS

Address: Meidlinger Hauptstraße, 1120 Vienna, Austria. **Completion:** ongoing. **Planning partners:** Zivilingenieur für Bauwesen Kurt Traxler. **Area:** 38,150 m². **Paving:** granite. **Landscaping:** plane grove. **Street furniture:** benches, seats, lighting. **Context:** shopping street.

↖ | **Site plan,** Meidlinger Platzl
↓ | **Fountain on Meidlinger Platzl**

← | Promenade under plane grove
↓ | Site plan

scape Landschaftsarchitekten

↑ I **Area accented with green lawns and trees**
↗ I **Wide pedestrian route with seating**
→ I **Various paved surfaces**

Pedestrian Zone

Kamen

The historic city center of Kamen – a city with 48,000 inhabitants, situated in the eastern part of the "Ruhrgebiet" – is a shopping and service location of more than local significance. The aim of the redesign was to give Kamen's heterogeneous center a new identity and unity. A strikingly new and timeless quality in form and material sets urban design accents and contributes to the positive image of the city. The three spatial sequences are defined according to their different characters. In contrast to the hard landscape of the shopping street, Adenauer Street has been transformed into a green boulevard with a promenade and recreation areas. Willy-Brandt-Square is the focal point within these spatial sequences. A paving strip forms a continuous carpet through the entire pedestrian zone, uniting the historical and modern architectural structure.

PROJECT FACTS

Address: Willy-Brandt-Platz, 59174 Kamen, Germany. **Completion:** 2008. **Length:** 1,400 m. **Area:** 22,000 m². **Estimated visitors:** 5,000 per day. **Paving:** natural paving stones of alternating yellow and gray granite. **Landscaping:** Fraxinus angustifolia 'Raywood', Platanus acerifolia. **Street furniture:** L. Michow & Sohn, scape. **Context:** transformation of historical city center.

← | Public square lined with trees
↓ | Landscaping is an integral part of this design

↑ | Site plan
← | Water feature

Gustafson Guthrie Nichol

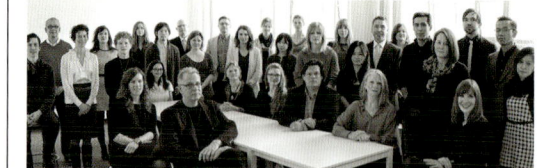

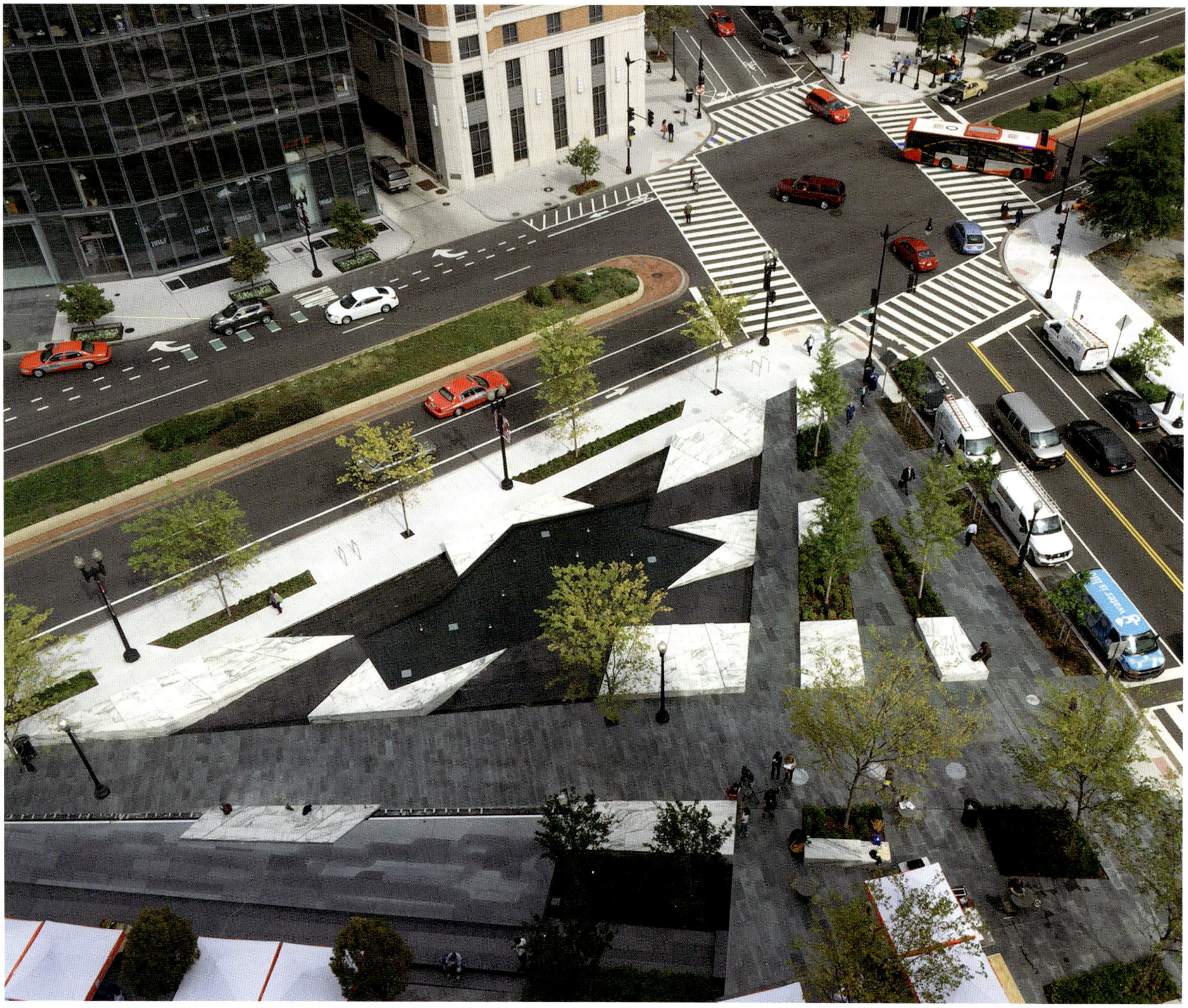

↑ | Park at CityCenterDC
↗ | Plaza at CityCenterDC
→ | Pedestrian route between buildings

CityCenterDC

Washington

CityCenterDC, in the heart of Washington DC and one of the largest downtown projects in the United States, is an urban center mixing commercial, retail, and residential uses. The landscape and streetscape design features a diverse network of new public open spaces, including pedestrian alleys, neighborhood streets, plazas, and a neighborhood park. At the upper levels, the new buildings incorporate multiple landscaped terraces and green roofs. The project re-introduces the original street grid of Washington's L'Enfant plan as a framework for a pedestrian-oriented, vibrant, 18-hour mixed-use neighborhood that connects into the surrounding city fabric and continues the revitalization of this sector of the city.

PROJECT FACTS

Address: Washington DC, USA. **Completion:** 2013. **Architects:** Foster + Partners, Shalom Baranes. **Area:** 40,469 m². **Landscaping:** streets, lanes, alleyways and gardens. **Street furniture:** integrated seating elements. **Context:** shopping and residential area in city center.

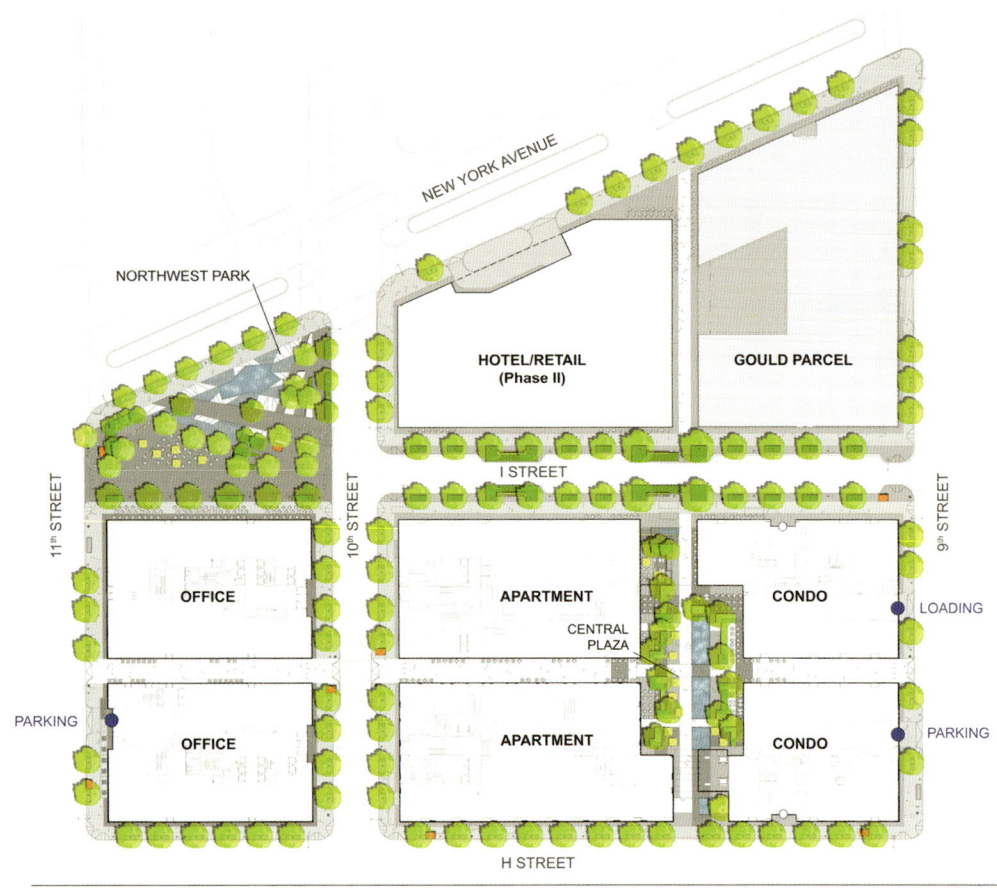

NEW YORK AVENUE
NORTHWEST PARK
HOTEL/RETAIL
(Phase II)
GOULD PARCEL
I STREET
11th STREET
10th STREET
9th STREET
OFFICE
APARTMENT
CONDO
LOADING
CENTRAL
PLAZA
PARKING
OFFICE
APARTMENT
CONDO
PARKING
H STREET

↖ | Site plan
↓ | Shopping street

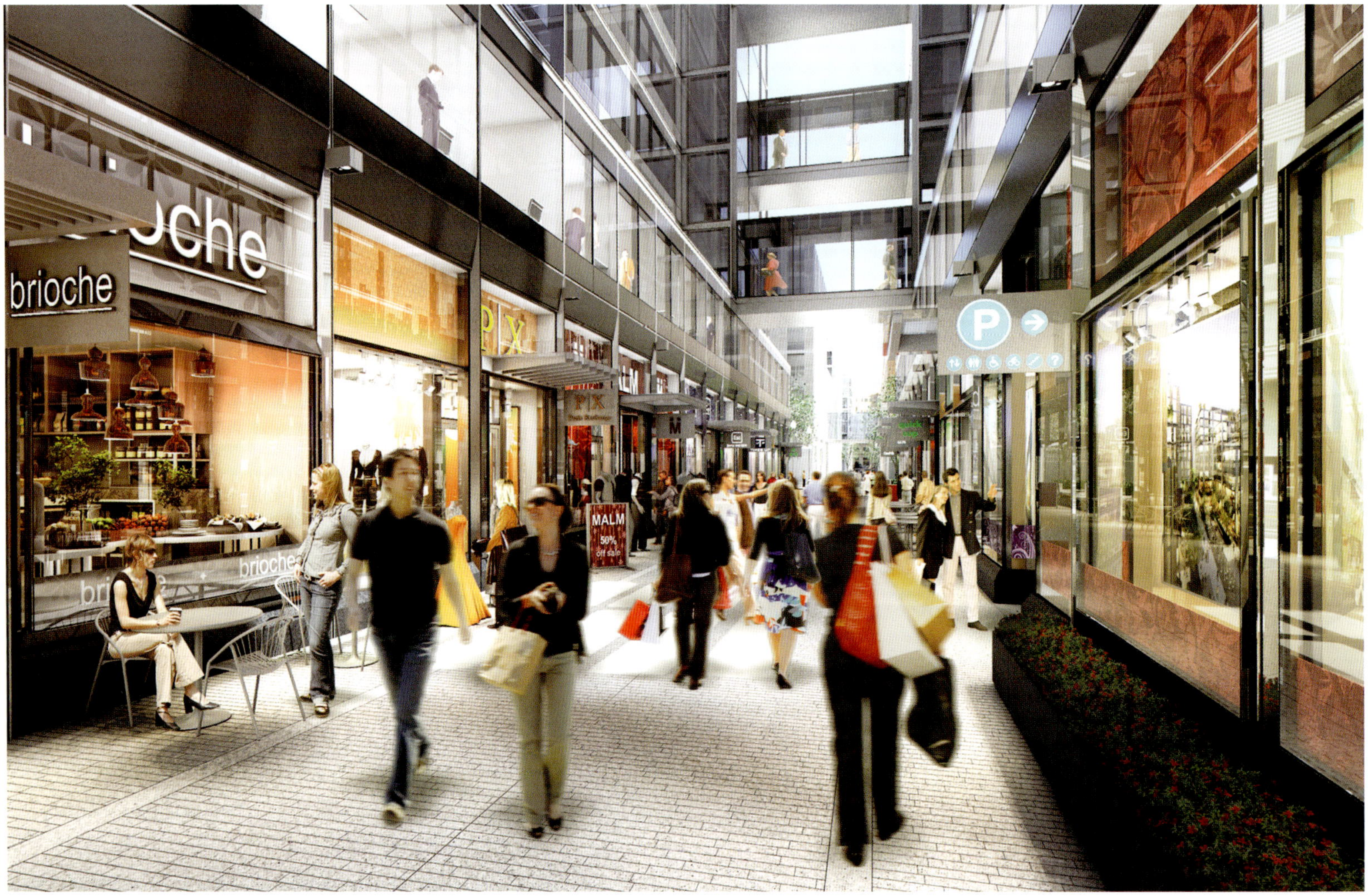
brioche
MALM
50%
off sale

↑ | **Design concept**
← | **Pedestrian zone,** bird's-eye view

↑ | **Street furniture,** seating accommodation
↓ | **Lively pedestrian zone**

→ | **View of Kirchgasse**
↙ | **Bronze inlay**
↓ | **Water course,** at "Schützenhofquelle"

Pedestrian Zone

Wiesbaden

The redesign of this pedestrian zone in Wiesbaden pays particular attention to the simplicity and elegance of the individual design elements. The pedestrian routes have been laid with light granite paving and in places are accentuated with shimmering bronze inlay. Various types of paving make the organization of the streets clearer, thus aiding orientation. Special features, such as simple reduced water elements or groups of trees fit into the context as a whole and give the shopping street its individual character and transforming it into a space where visitors are not only encouraged to spend time, but actually want to. Elegant lighting columns and suspended lights bathe the streets in a warm light.

PROJECT FACTS

Address: Kirchgasse und Langgasse, 65183 Wiesbaden, Germany. **Completion:** 2009. **Length:** 850 m. **Area:** 30,000 m². **Estimated visitors:** 11,070 per hour. **Paving:** asphalt, granite, basalt, bronze inlay. **Landscaping:** 25 plane trees. **Street furniture:** benches designed by ST raum a., stones from BESCO GmbH, streetlights from Philips. **Context:** shopping street in city center.

← | **View of Mauritiusplatz**
↓ | **Site plan,** Kirchgasse and Langgasse

↖ | Street compartmentation
↑ | Fountain
← | Entrance to Langgasse

Valentien + Valentien

![Historic street scene in Ochsenfurt]

↑ | **Historic buildings**, with "Ochsenfurt"
↗ | **Main street**, after conversion
→ | **Street furniture**, at town hall

Historic City Redevelopment

Ochsenfurt

The market square is the central shopping area in the town of Ochsenfurt. The construction of a new road to the south of the town created the right conditions to enable renovation and improvement works to take place in the Old Town. In order to improve the historical ensemble for residents and visitors alike, the area in front of the town hall was landscaped with trees and benches. New paving has also been added, which helps to improve the appearance of the town. The fountain on the southerly market square comprises a long water channel lined with slate and featuring a stone statue of an ox crossing a ford. This is a reference to the town's name Ochsenfurt (ox ford).

PROJECT FACTS

Address: Hauptstraße, 97199 Ochsenfurt, Germany. **Completion:** 2007. **Length:** 500 m. **Area:** 4,850 m². **Paving:** granite. **Landscaping:** plantings, five trees, four hornbeams. **Street furniture:** benches, sunshades near the restaurants, steel bike stands, fountain. **Context:** redevelopment of historic city center.

← | **Statue of St. George,** new church wall
↓ | **Site plan**

↑ | View of town hall
← | Town hall forecourt

↑ | **Bird's-eye view,** restored fountain
↓ | **Rectangular paving pattern**
↘ | **Detail fountain,** child playing

↗ | **Meeting point**
→ | **Hugo-Bork-Platz,** fountain
↓ | **City parquet**

Porschestraße Pedestrian Zone

Wolfsburg

Originally Porschestraße was designed by the planners to be a wide, open street for traffic and retail that connected the train station with the south of the city. It was transformed into a pedestrian zone in the 1970s with numerous shopping and gastronomy pavilions. This new design now unites the advantages of both concepts: An open spatial arrangement paired with leisure elements and retail. The new paving is of light natural stone and structures Porschestraße into five functional strips. Plane trees have been planted on the middle strip, with benches between the trees that offer a space where one can simply sit and watch the world go by. The extended commercial spaces are protected by projecting glass roofs, making shopping a comfortable experience whatever the weather.

PROJECT FACTS

Address: Porschestraße, 38440 Wolfsburg, Germany. **Completion:** 2009. **Lighting designers:** Fahlke & Dettmer. **Architects cantilevered roofs:** O.M. Architekten. **Length:** 1,300 m. **Area:** 36,500 m². **Estimated visitors:** 4,245 per hour. **Paving:** 84 % granit, 16 % steel. **Landscaping:** 50 plane trees. **Street furniture:** benches designed by ST raum a., stones from BESCO GmbH, built from Elancia. **Context:** main shopping street in city center.

← | **Strolling under plane trees**
↓ | **Plan,** northern and middle section

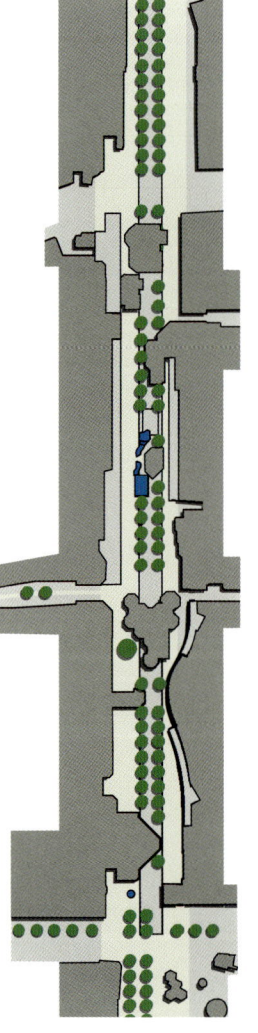

↖ | **Fountain,** seating cubes
↑ | **Round bench,** tree bed
← | **Pavement structure**

↑ | **Bagby Street,** pedestrian zone
↗ | **Babgy Street at night**
→ | **Gallon,** rain garden

Bagby Street Reconstruction

Houston

Bagby Street reconstruction in Houston's Midtown District is a renewed effort by the district to connect with current demographics, modern infrastructure demands and long-term sustainability goals. It envisioned a highly programmed, pedestrian friendly environment encouraging private re-investment, and a high level of service for current residents. Goals relate to limiting disturbance to local business, the district's return on investment, sustainability features, human comfort, and reduction of maintenance burden. A master planning and analysis approach ensured the project matched current land uses, related to future redevelopment opportunities, and was responsive to local demographics.

PROJECT FACTS   **Address:** Midtown District, Houston, TX, USA. **Completion:** 2013. **Length:** 997 m. **Planning partners:** Walter P. Moore. **Area:** 22,864 m². **Paving:** pavestone, stepstone. **Context:** mixed-use corridor.

← | **Plantings**
↓ | **Site plan**

↑ | **Rain garden**
← | **Bicycle rack**

Aspect Studios

↑ | **Custom-designed furniture**

↙→ | **Artwork installation,** "Between Two Worlds" by Jason Wing

Chinatown Public Domain Upgrades

Sydney

The upgrades were the first phase of remodeling work undertaken on Chinatown's public domain. The focus of the work is to improve the quality of the space and strengthen pedestrian areas. Aspect Studios' scope included rectification work to the art installation "Heaven and Earth", located at the intersection of Little Hay and Dixon Street. Factory Street has been transformed with a fine grain combination of granite pavers and cobbles. Mature trees have been planted in Little Hay Street and Factory Street as well as an overlay of new lightboxes. A site specific ribbon of custom-designed furniture completes the works assisting to reactivate these streets. Aspect Studios worked closely with artist Jason Wing. The brightly colored work includes painted elements, granite etching and site-specific lighting.

PROJECT FACTS

Address: Little Hay Street, Kimber Lane and Factory Street, 2000 Sydney, Australia. **Completion:** 2012.
Artists: Jason Wing, Peter McGregor. **Area:** 2,790 m². **Paving:** granite paving, cobbles. **Landscaping:**
Robihia pseudoacacia 'Frisia'. **Street furniture:** steel framed custom-designed benches with timber tops
and backrest. **Context:** redeveloped inner city pedestrian zone.

← | Seating in retail area

↑ | **Art installation,** "Heaven and Earth" by
Peter McGregor and Deuce Design

↑ | **Art installation,** "Between Two Worlds" by
Jason Wing
↓ | **Site plan,** Factory Street

RETHINK

↑ | **Red Square,** overview
→ | **Bird's-eye view,** from Black Market to Red Square

Superkilen

Copenhagen

Superkilen is a heterogeneous site-collage in a dense, centrally located neighborhood in Copenhagen. The strongly international quarter with a mix of different cultures has been revitalized using open space as a physical framework. The concept aims at enhancing the diverse characters of its protagonists and within the site. The design reattributes an essential motif from garden history. In the garden, the translocation of an ideal, the reproduction of another place, of a far off landscape, is a common theme through time. The furnishings are developed from an international catalogue of urban design elements. In many months of workshops and conversations with residents and local associations the creativity and fantasy of the quarter has been mobilized. Civic participation has been developed as a motor for the design principle.

PROJECT FACTS

Address: Nørrebrogade, 2200 Copenhagen, Denmark. **Completion:** 2012. **Architects:** BIG Architects. **Artists:** Superflex. **Length:** 750 m. **Area:** 27,000 m². **Paving:** Polyurethane on asphalt, tartan coating, asphalt. **Landscaping:** ten different types of tree species. **Street furniture:** more than 100 different objects from more than 50 different countries. **Context:** urban park project in multi-cultural context.

← | **Iraqi swing set benches**
↓ | **Japanese octopus,** bird's-eye view

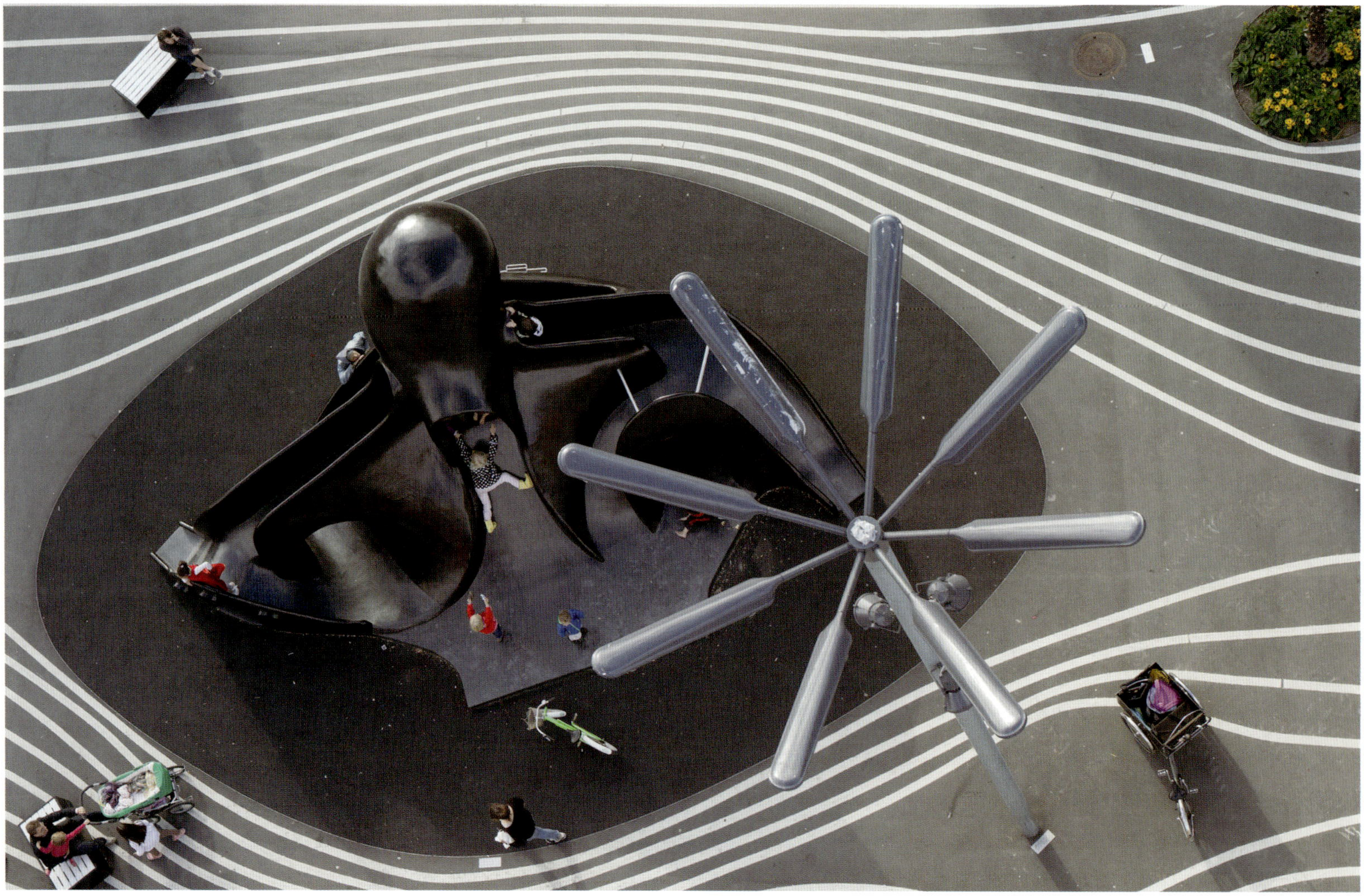

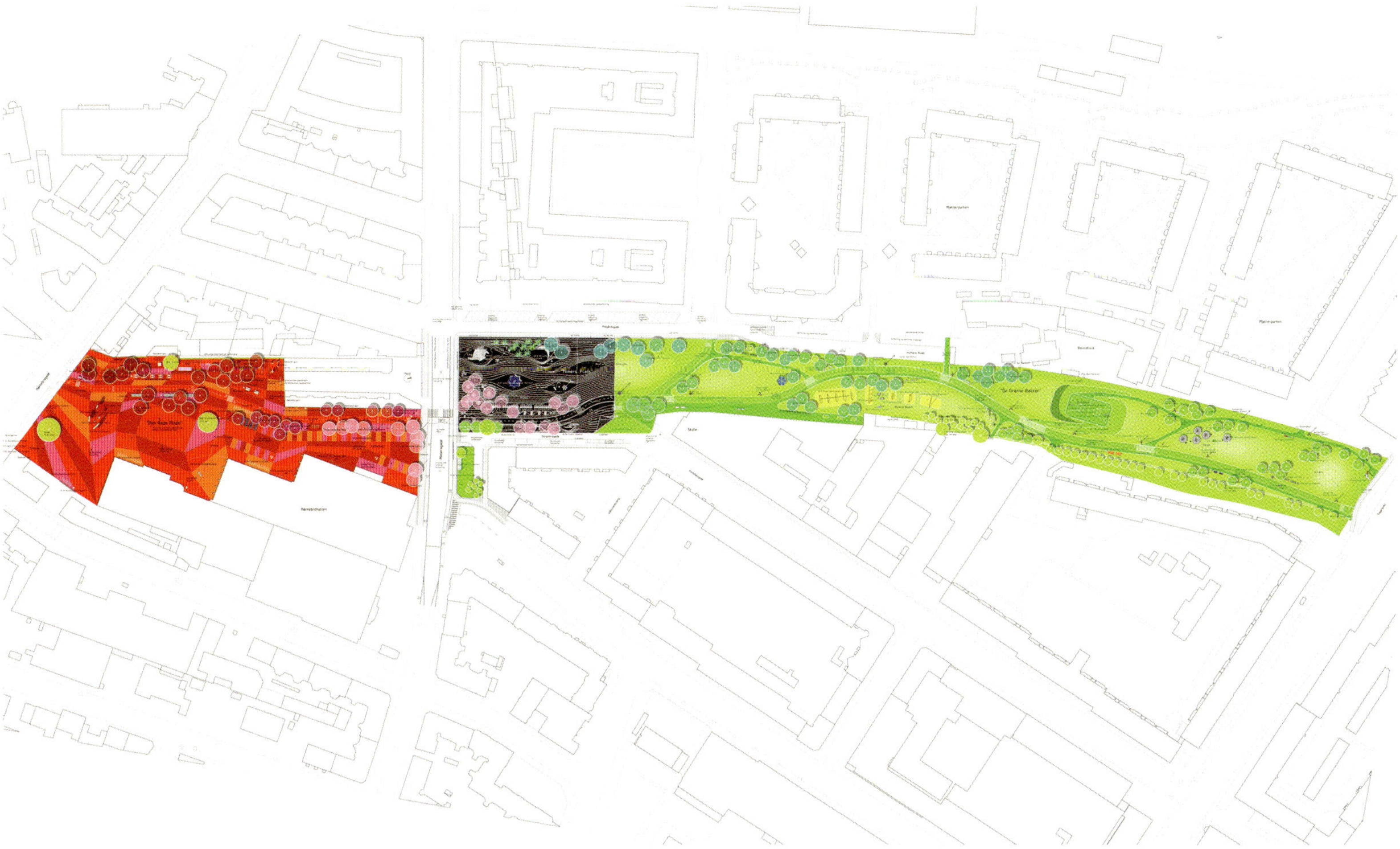

↑ | **Site plan,** Red Square, Black Market and Green Park
← | **Black Market,** evening

↑ | **Funnel interior**
→ | **Funnel-shaped column,** supports membrane roof

Mega Structure Expo 2010

Shanghai

The central boulevard is an emblem of the Shanghai Expo 2010. This 1000-meter-long and 100-meter-wide axis guides visitors to SBA exhibition pavilions. The boulevard serves as the central entrance and offers around 350,000 square meters of useable space. The axis is covered with the world's largest membrane. The roof is supported by 19 interior and 31 exterior masts, as well as six funnel-shaped mesh columns of steel and glass that are 45 meters in height and project upto 80 meters outward. These guide the daylight into the lower levels, giving them their name "Sun Valleys". The structure is one of five that will remain after the exhibition and in the long term it will form the center of a new city district.

PROJECT FACTS   **Address:** Expo site, Shanghai, China. **Completion:** 2010. **Area:** 5,280,000 m². **Context:** entrance boulevard to exhibition.

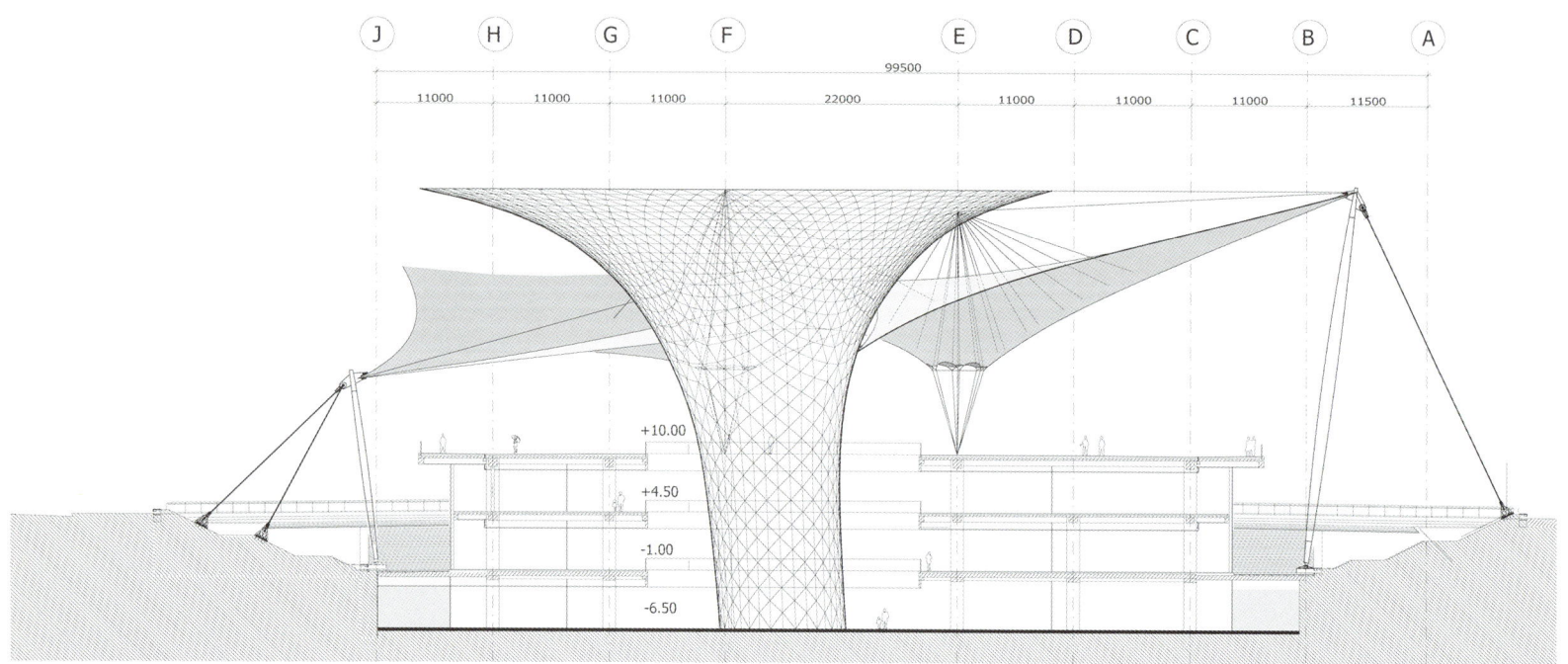

↑ | **Section**
↓ | **Mesh and membrane construction,** light
shines through

↓ | **Funnels,** draw daylight into lower levels

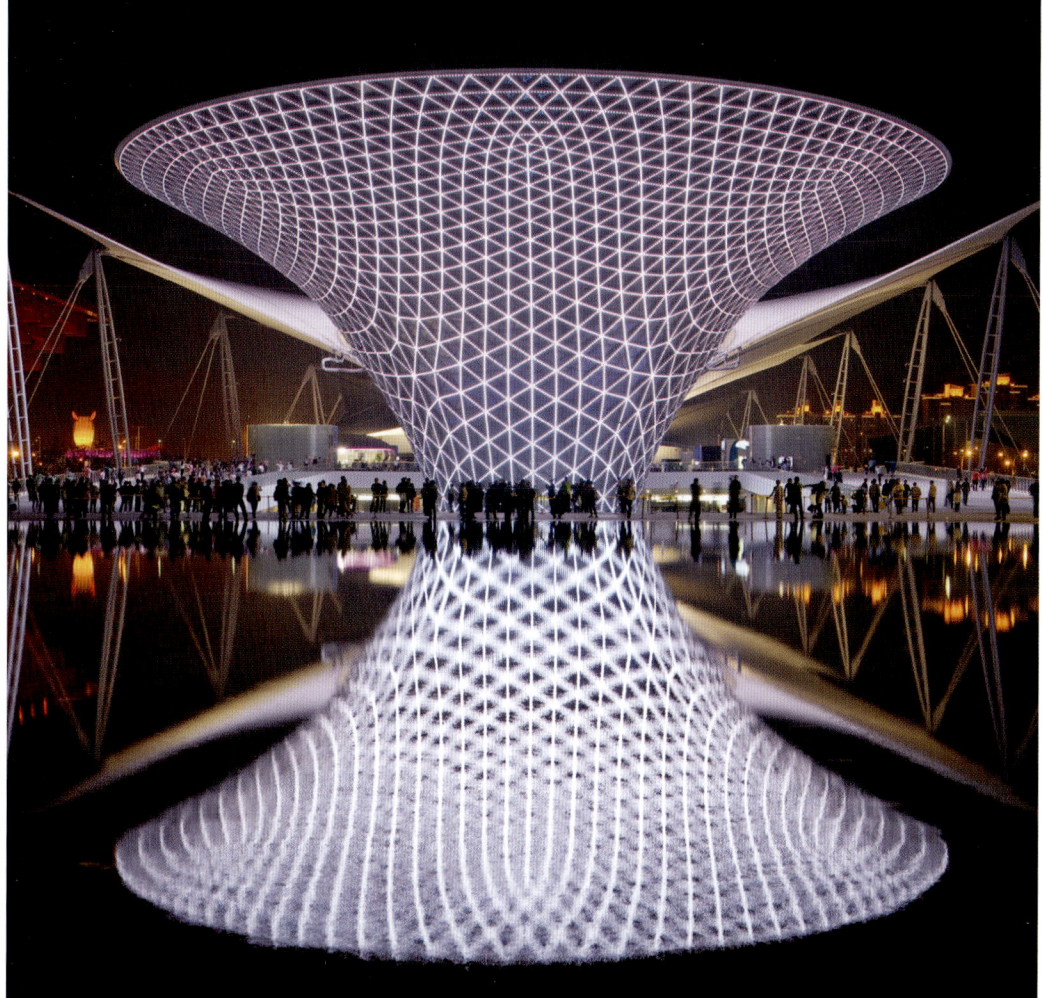

↑ | **Pedestrian boulevard,** partially covered
with membrane roof
← | **Illuminated funnel**

Claude Cormier et Associés

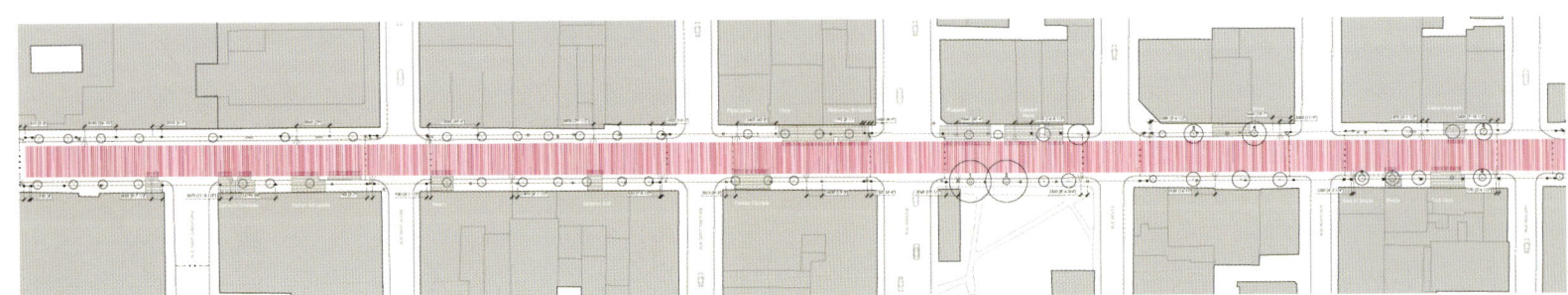

↑ | **Pedestrian area,** shopping street
↓ | **Site plan,** suspended pink ribbon

Pink Balls

Montreal

Pink Balls is a one-kilometer-long ribbon-like installation of 170,000 resin balls suspended over Sainte-Catherine Street East, in Montreal's gay village. It is part of the pedestrian transformation of the street into a pedestrian mall over the summer. The plastic balls, in three different sizes and five shades of pink, are strung together with wire, crisscrossing the street and stretching through tree branches at varying heights. The installation has been deployed in nine sections, each section displaying its own pattern. By integrating the community into the project, Pink Balls has increased the appeal among both local visitors and tourists, becoming a catalyst for economic and social development in a neighborhood that has otherwise been gripped by serious social and economic problems.

PROJECT FACTS

Address: Sainte-Catherine Street East, Montreal H2S 2B1, Canada. **Completion:** 2014. **Planning partners:** Société de développement commercial du village / Aires Libres. **Length:** 1,000 m. **Area:** 8,250 m². **Paving:** asphalt. **Landscaping:** existing street trees. **Street furniture:** existing street furniture. **Context:** pedestrian mall during summertime, installation.

↑ | **Pedestrian mall**
↓ | **Aerial view from Jacques-Cartier Bridge**

↓ | **Pink Balls,** shadows on the ground

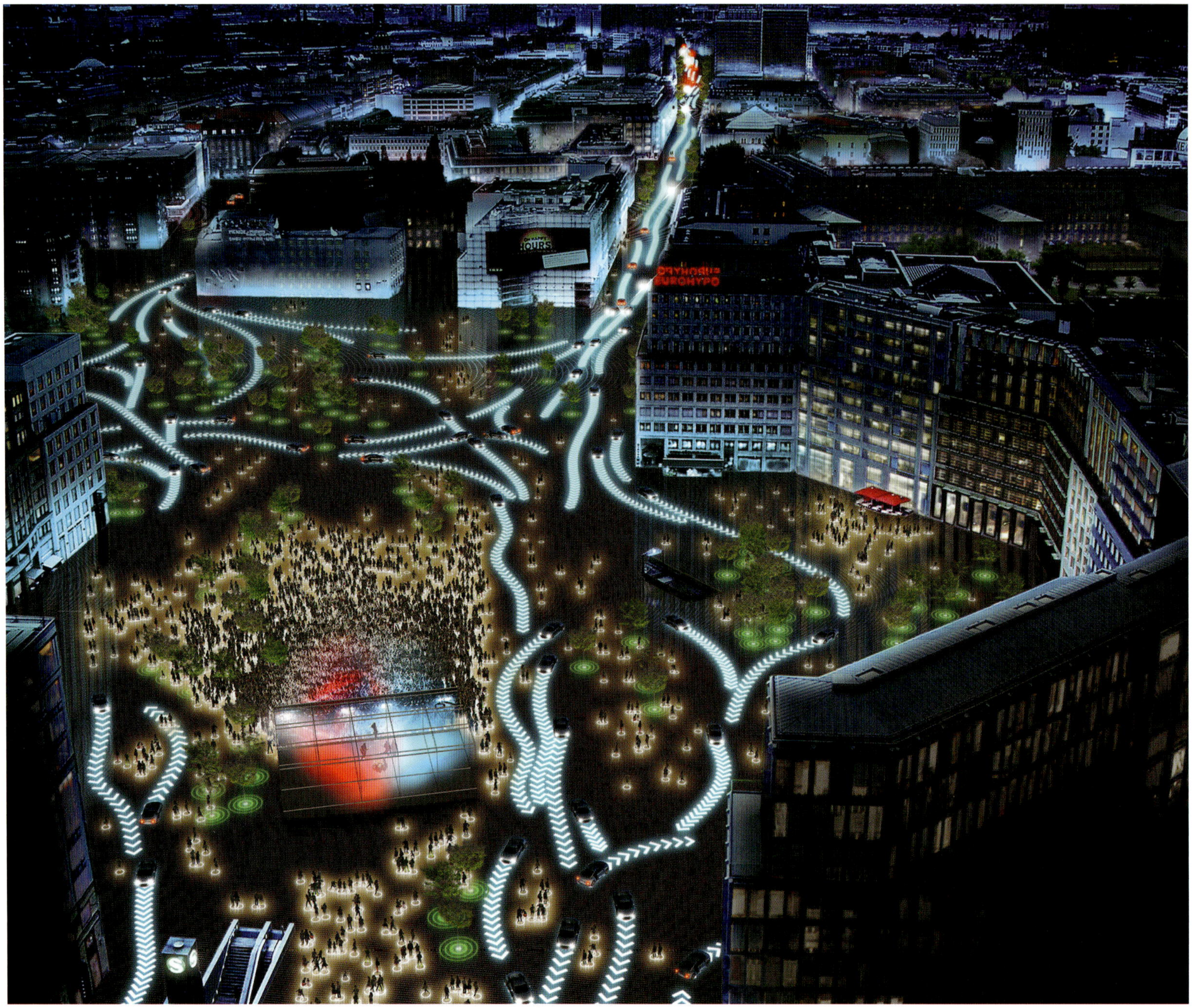

↑ | **Concept,** from above
↗ | **Circles demarcate 'safe zones'**
→ | **Concept at night**

Audi Urban Future

Berlin

The futurist and prolific inventor Ray Kurzweil describes evolution in technology as an exponential curve following the law of accelerating returns. Technological breakthroughs are not only happening constantly but they are happening faster and faster. Could it be that the next revolution of urban space will be caused not by flying cars or advances in speed and engine power, but by the full merger of information technology with personal transport? The driverless car; the first truly auto-mobility vehicle; could be the next revolution in urban space. Picture a city in 25 years where vertical façades appear unchanged, but the city pavement is transformed into a reprogrammable surface replacing the fixed elements of driveway, sidewalk or square; a digital street surface completely re-animating and re-organizing a familiar city.

PROJECT FACTS

Address: undefined. **Design completion:** 2010. **Area:** undefined. **Paving:** three-dimensional LED. **Context:** traffic concept, installation for Design/Miami 2011.

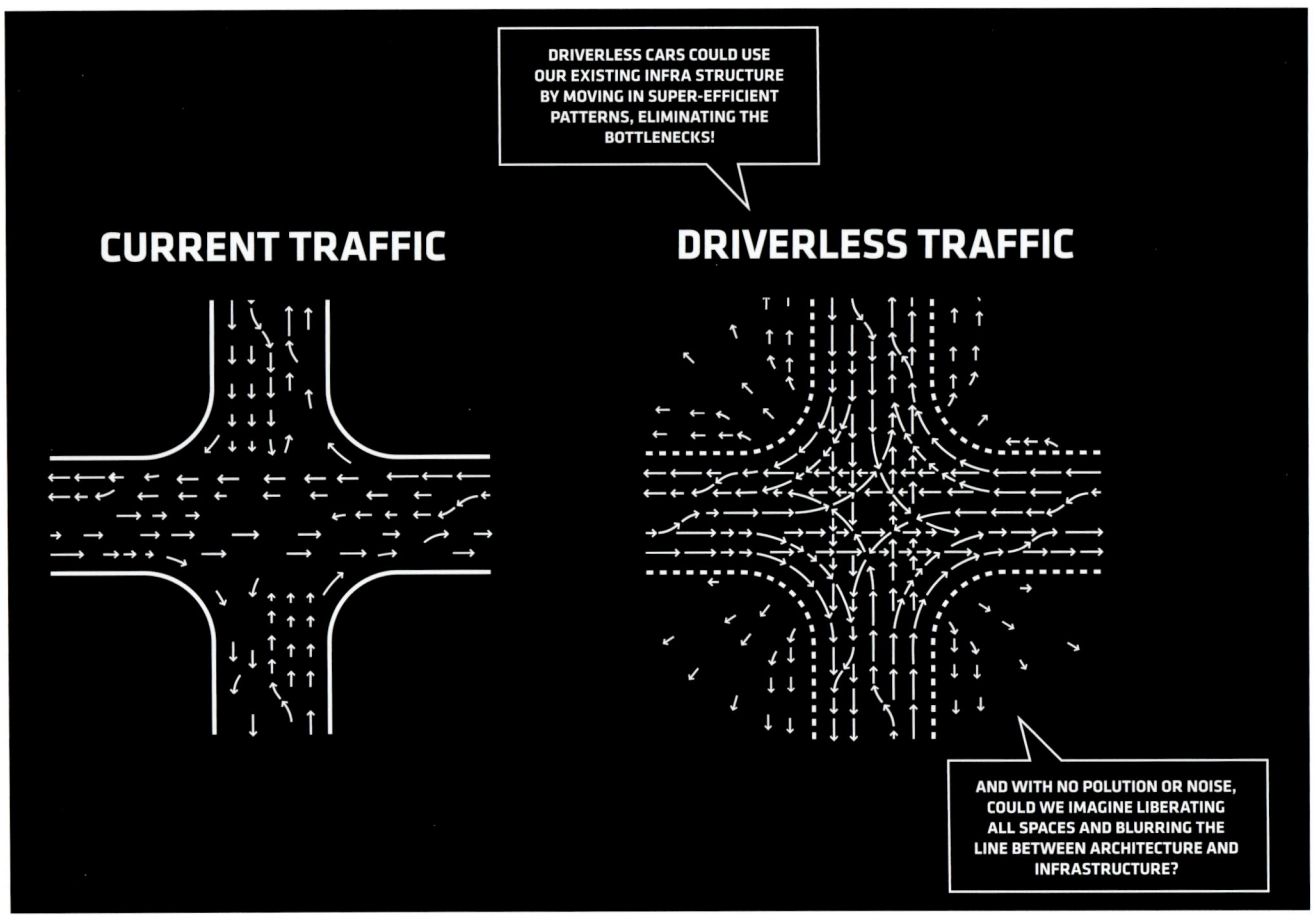

↑ | Concept, current and driverless traffic
↓ | Audi exhibition

↑ | Concept integrated into city
↓ | Illuminated circles show pedestrian 'safe zones'

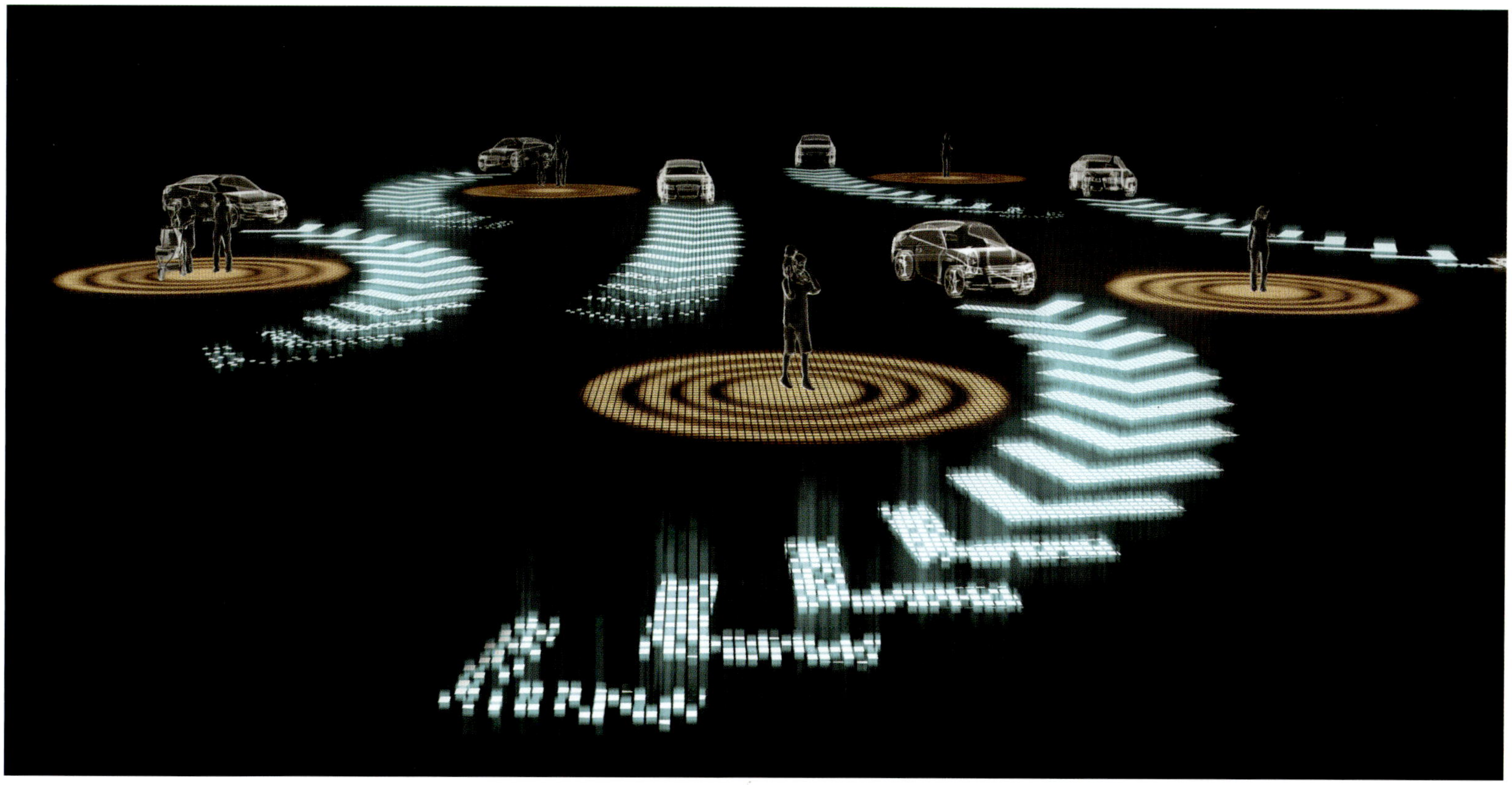

↑ | Wesley Quarter
↗ | Historic Wesley Church
→ | Illuminated podium

Wesley Quarter

Perth

The Wesley Quarter renewal has reinvigorated the public spaces and surrounding features of Perth's historic Wesley Church. The city precinct fronts a vibrant fashion center, calling for an elegant design response that harmonizes with both the contemporary and heritage architecture present. A monochromatic palette of natural charcoal granite and white sandstone subtly defines and connects the seating areas, transitional spaces and pedestrian thoroughfares. A typographical narrative of the site's history unfolds along the contrasting laneway pavers. A raised stone podium and tree plinth provide platforms for activation and contemplation. They are lit from beneath, complementing the church façade – part of a strategy to illuminate Perth's built heritage as a rich component of the contemporary urban fabric.

PROJECT FACTS

Address: Corner of Hay and William Streets, 6000 Australia. **Completion:** 2009. **Area:** 100 m². **Paving:** natural charcoal granite and white sandstone. Etched text filled with resin. **Context:** mixed-use urban streetscape with heritage interface.

← | **Interpretative laneway pavers**
↓ | **Detail plan,** laneway and podium

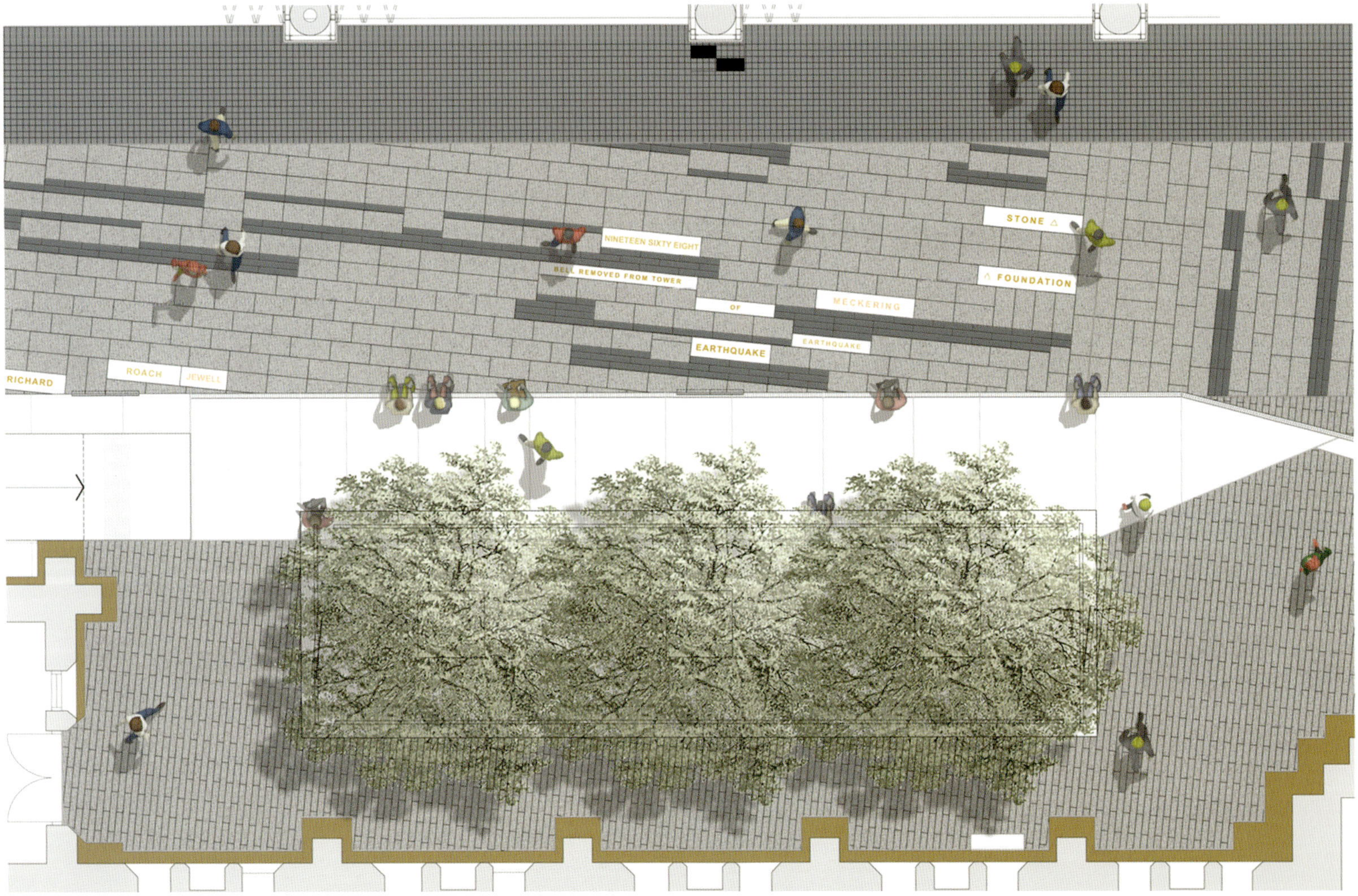

↑ | Pedestrian zone at night
← | Podium

AS&P –
Albert Speer & Partner

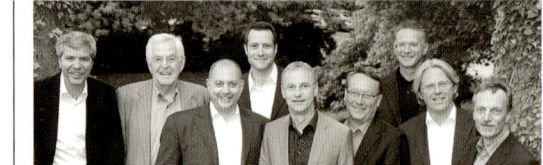

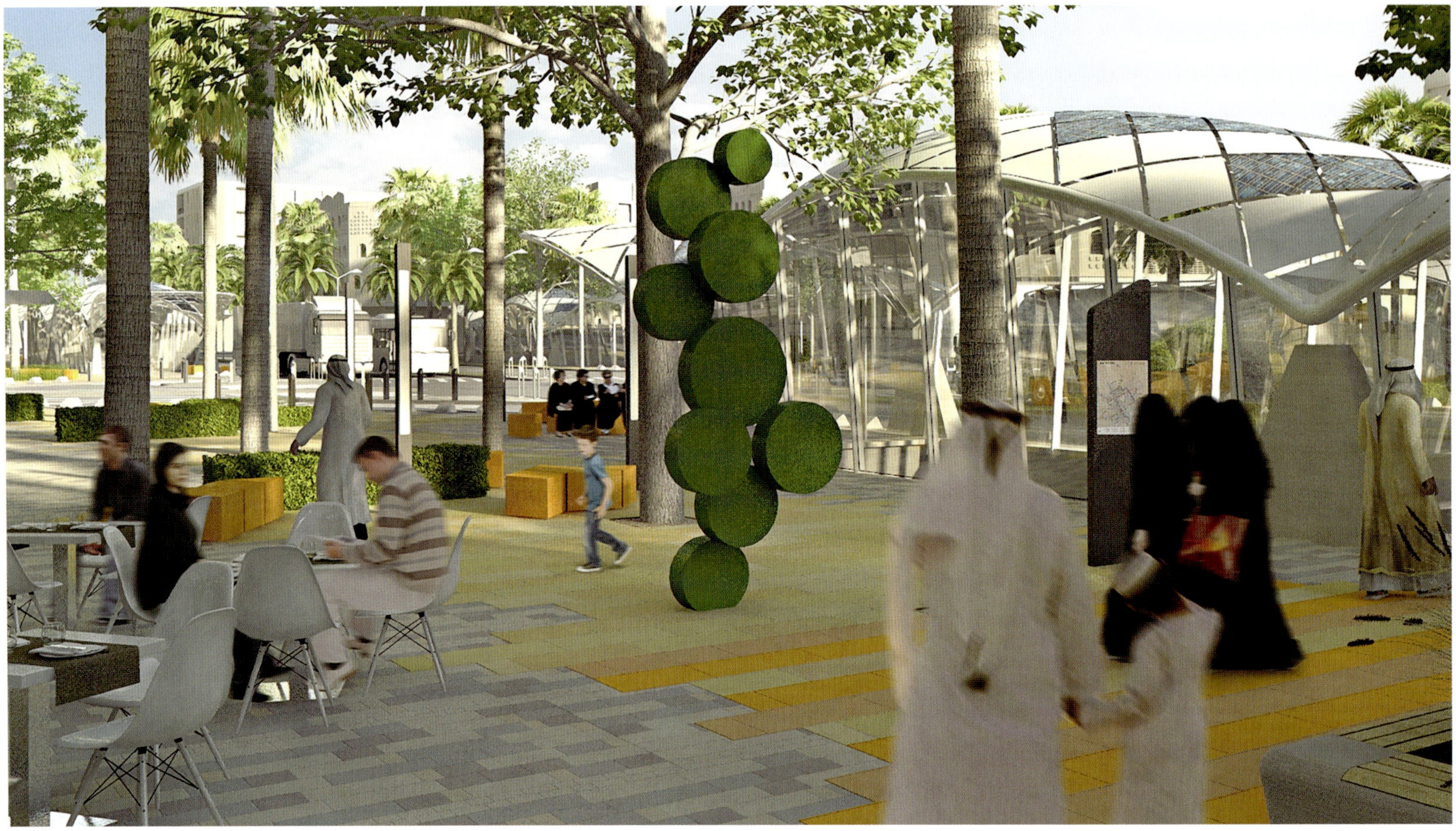

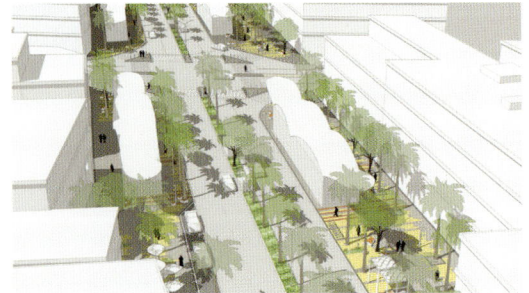

↑ | **Inner city pedestrian area,** metro station

↙ | **Typical streetscape design**
↓ | **Site plan**

Metro Urban Design and Streetscape

Riyadh

A comprehensive public transport network (metro and express bus) is planned for the metropolitan city of Riyadh, and is viewed as an historical chance to improve the quality of, and access to, the traffic-heavy and largely unesthetic streetscape for pedestrians and cyclists. In this context, the Arriyadh Development Authority commissioned AS&P to develop an Urban Design and Streetscape Manual (UDM), which specifies functional and design guidelines for general planners. The UDM includes typified designs for the public space along the metro corridor, a design matrix of design elements, and a flexible design theme, which ensures high quality and simple implementation.

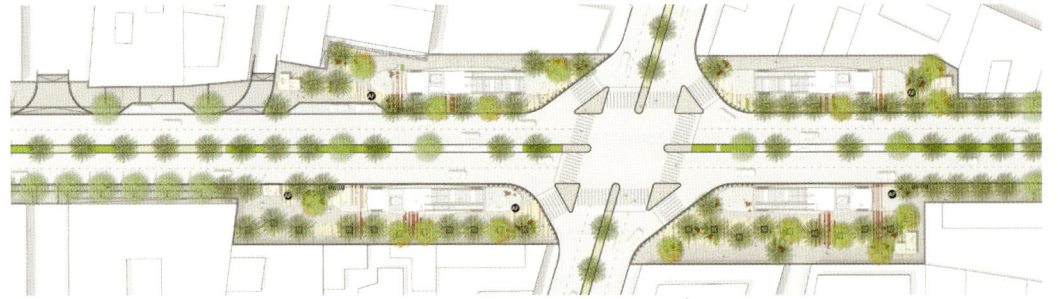

PROJECT FACTS   **Address:** Riyadh, Saudi Arabia. **Completion:** ongoing. **Length:** 90,000 m. **Area:** 3,600,000 m². **Paving:** mix of high quality pre-cast and natural materials. **Landscaping:** rich mix of plants and trees including indigenous species. **Context:** Riyadh Public Transport Program.

↑ | **Residential zone,** mixed-use area
↓ | **Residential context of metro passage**

↑ | City space 'Musikhuspassagen'

Thomas B. Thriges Gade

Odense

Thomas B. Thriges Gade is the name of a street that was constructed in the 1960s to modernize Odense, and make room for the increasing traffic. The highway plowed a separating trail through the old alleys to ensure accessibility. It is on this asphalt road that a lush and visionary city will arise. A lush green identity will give the new quarter a strong and unified appearance. It takes density to hold vibrant city life, and from that firm conviction, liveliness is cared for all the way from the paving to the functions of the ground floor, the materials and rhythmic changes of the streetscape, safety and lighting, variations from vibrant to quiet places, from regional to neighborhood spaces. A new light rail of sustainable transport will stretch through the area and tie Odense closer together.

PROJECT FACTS

Address: Thomas B. Thriges Gade, 5000 Odense, Denmark. **Completion:** 2020. **Length:** 700 m. **Area:** 51,000 m². **Paving:** bricks. **Context:** new quarter in town center.

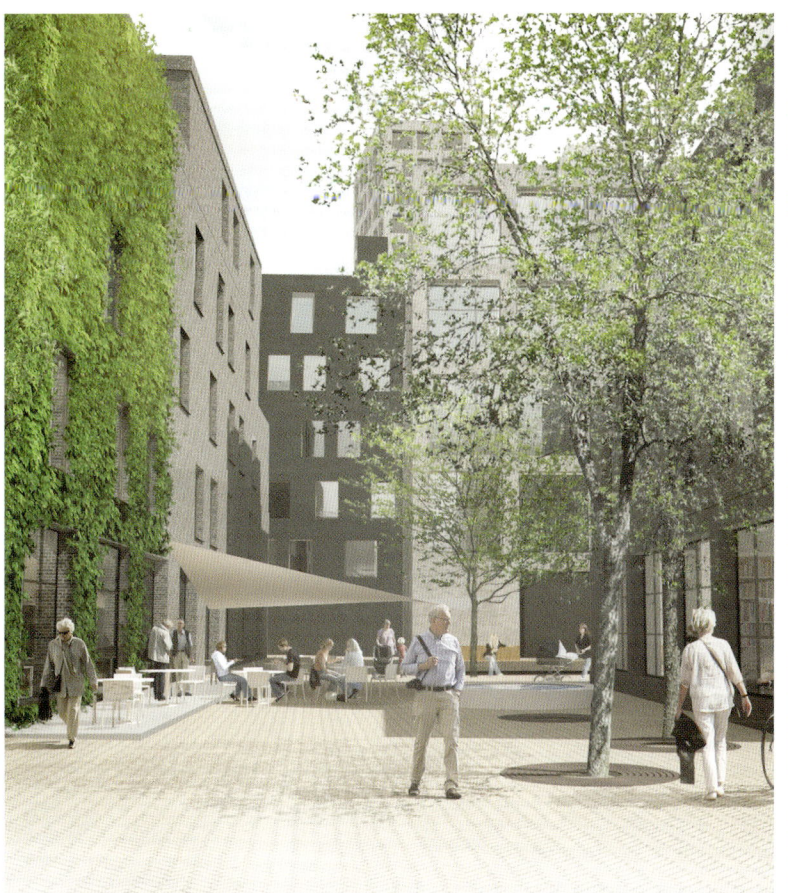

↑ | **Frokostpladsen**

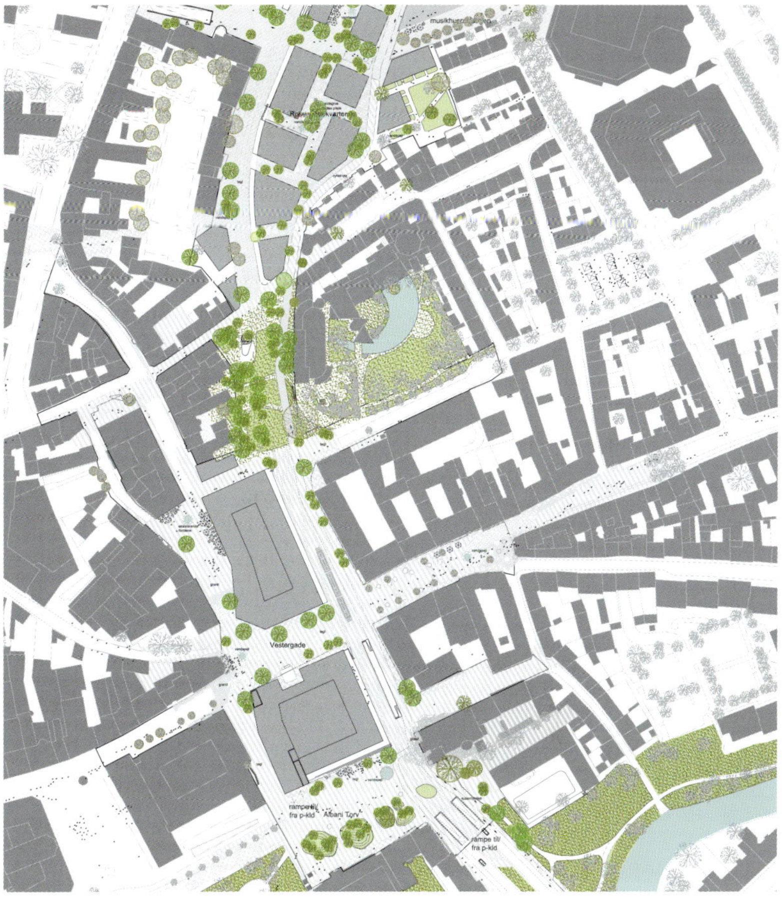

↑ | **Master plan**
↓ | **Central green space,** light rail tracks

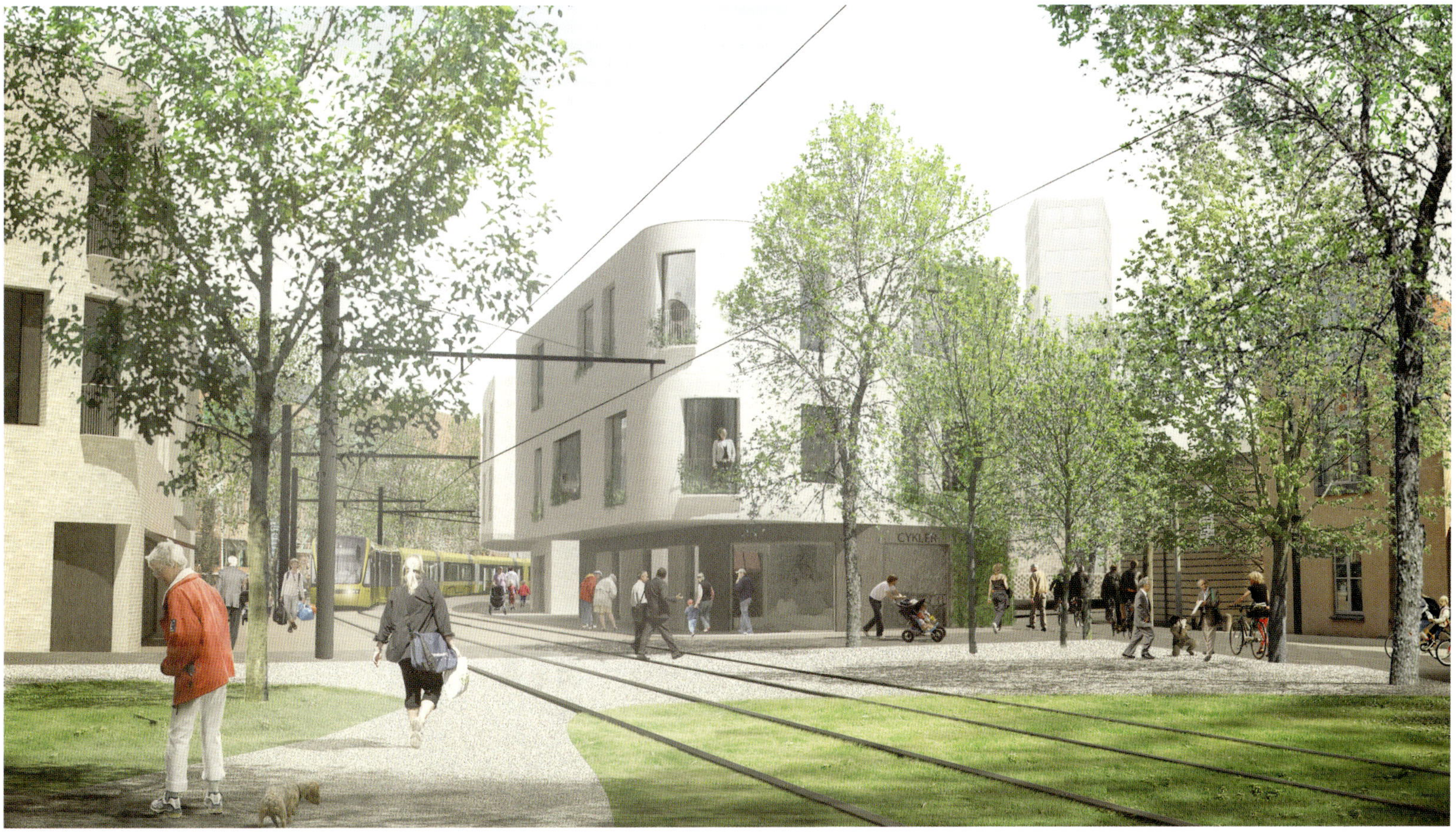

Moore Ruble Yudell
Architects and Planners

↑ | **Shaded Central Plaza**
→ | **Interactive fountain,** sun screen

Master Plan and Town Center

Camana Bay

Camana Bay is a privately developed new town establishing a fresh paradigm for development on Grand Cayman. The master plan envisioned a four village scheme offering a full range of amenities woven together by lushly landscaped pedestrian and bicycle paths, courtyards, canals, and vehicular streets of varying sizes. Located at the heart is a town center which is fronted by shops, a cinema, offices, restaurants and residential balconies. Simple, compact structures are woven together with the pedestrian oriented Paseo, Crescent and courts of a variety of sizes and characters shaped to encourage the prevailing cooling breezes and temper the strong, Caribbean sun. Created is a sense of lively community – authentic and contemporary grown with respect to the history, culture and climate of the place.

PROJECT FACTS

Address: Camana Bay, Grand Cayman KY1-1206, Cayman Islands. **Completion:** 2011. **Executive architects:** Spillis Candela (now AECOM). **Local architects:** The Burns Conolly Group. **Landscape architects:** Olin. **Length:** 227 m. **Area:** 3,690 m². **Estimated visitors:** 400 per day. **Paving:** concrete. **Landscaping:** Caribbean native plant materials. **Street furniture:** custom teak benches, custom resin internally lit 'bubble benches'. **Context:** new town center on island.

← | **Site plan,** town center
↓ | **The Paseo,** the town's center of activity

↑ | **Colorful bubble benches,** cinema courtyard
← | **The Paseo,** local rock and native plantings

↑ | **Plaza,** encourages collaboration and entrepreneurial activity
↗ | **Circuit used for exercising**
→ | **Informal destinations for meeting and socializing**

TechTown District

Detroit

TechTown – an emerging district in Midtown Detroit – is currently characterized by surface parking, vacant properties, and inward-facing hubs of activity. The TechTown District Plan articulates an inspiring vision for the revitalization of the district. Developed by Sasaki Associates in collaboration with Midtown Detroit Inc. and U3 Ventures, the plan accelerates innovation, promotes entrepreneurship, and builds community around the generation of ideas in a vibrant, mixed-use setting. Leveraging the potential of key institutional anchors within the district, the plan creates an environment that fosters knowledge generation and innovation.

PROJECT FACTS

Address: 440 Burroughs Street, Detroit, MI 48202, USA. **Completion:** 2013. **Urban design and local planning:** Interface-Studio. **Area:** 602,982 m². **Context:** knowledge district in midtown Detroit.

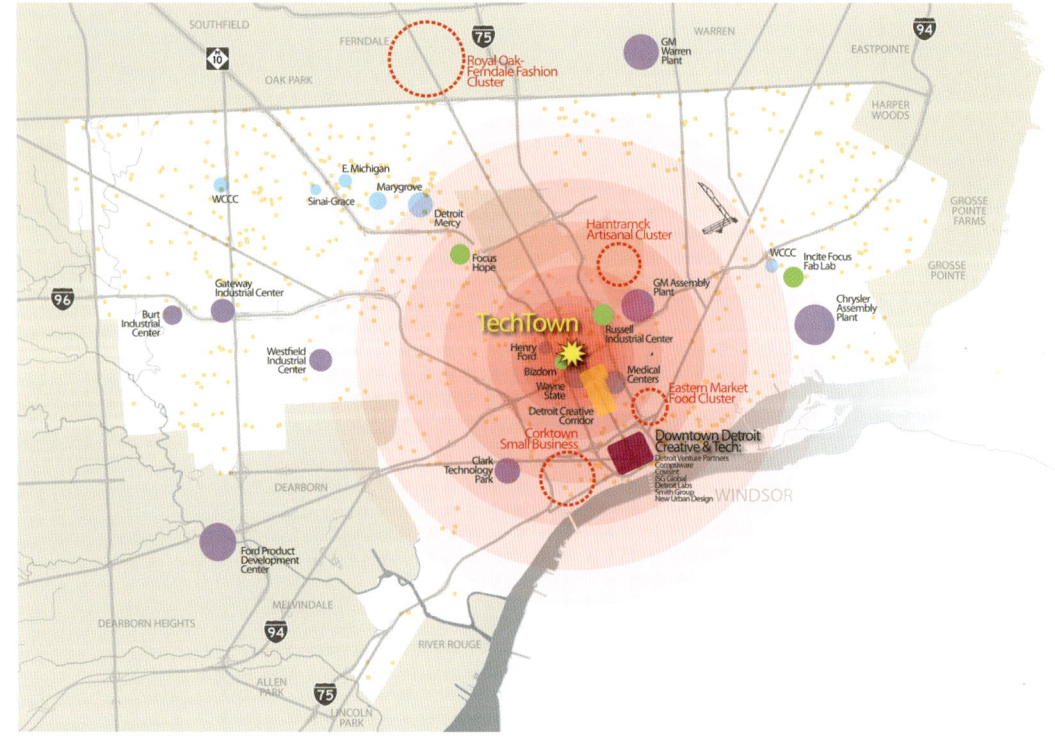

↖ | **Concept**
↓ | **Plaza,** designed for flexibility and to support
year-round activity

↑ | **Fabrication tables for testing and developing new products**
↙ | **Site plan**

↑ | **The promenade,** play of light and shadow
↓ | **Plan**

Promenade of Light

London

The Promenade of Light won the architectural competition launched by the Architectural Foundation in 2002. The brief was for the improvement of a grassed area in front of the shops, with strategic proposals for the areas surrounding the Old Street Roundabout. The existing 21 mature plane trees suggested a promenade that could be reinforced by adding more trees. 18 new trees were added, and a raised stone promenade was laid between the two rows of trees. Commuters, shoppers, couriers, school children, elderly, office workers, families, each emerges at different hours of the day, to inhabit different sections of the promenade. 23 lamp poles, each with six to eight lamps, illuminate the space from above.

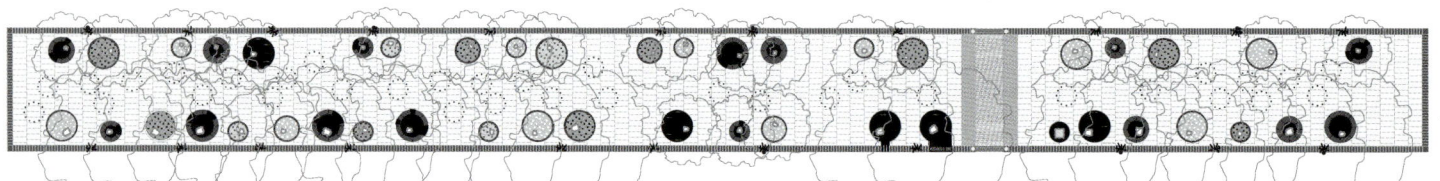
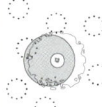

PROJECT FACTS

Address: Old Street, London EC1Y 1BE, United Kingdom. **Completion:** 2006. **Area:** 3,455 m². **Paving:** granite. **Landscaping:** trees and flowers in street planters. **Street furniture:** integrated seating elements. **Context:** urban street in city center.

↑ | **Lamp pole,** with individually positioned lamps

↑ | **Promenade of Light,** daytime
↓ | **Atmospheric light,** trees and flower beds

OKRA landschapsarchitecten

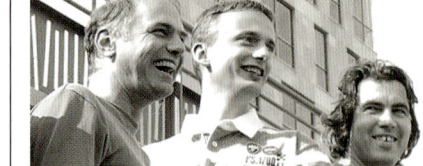

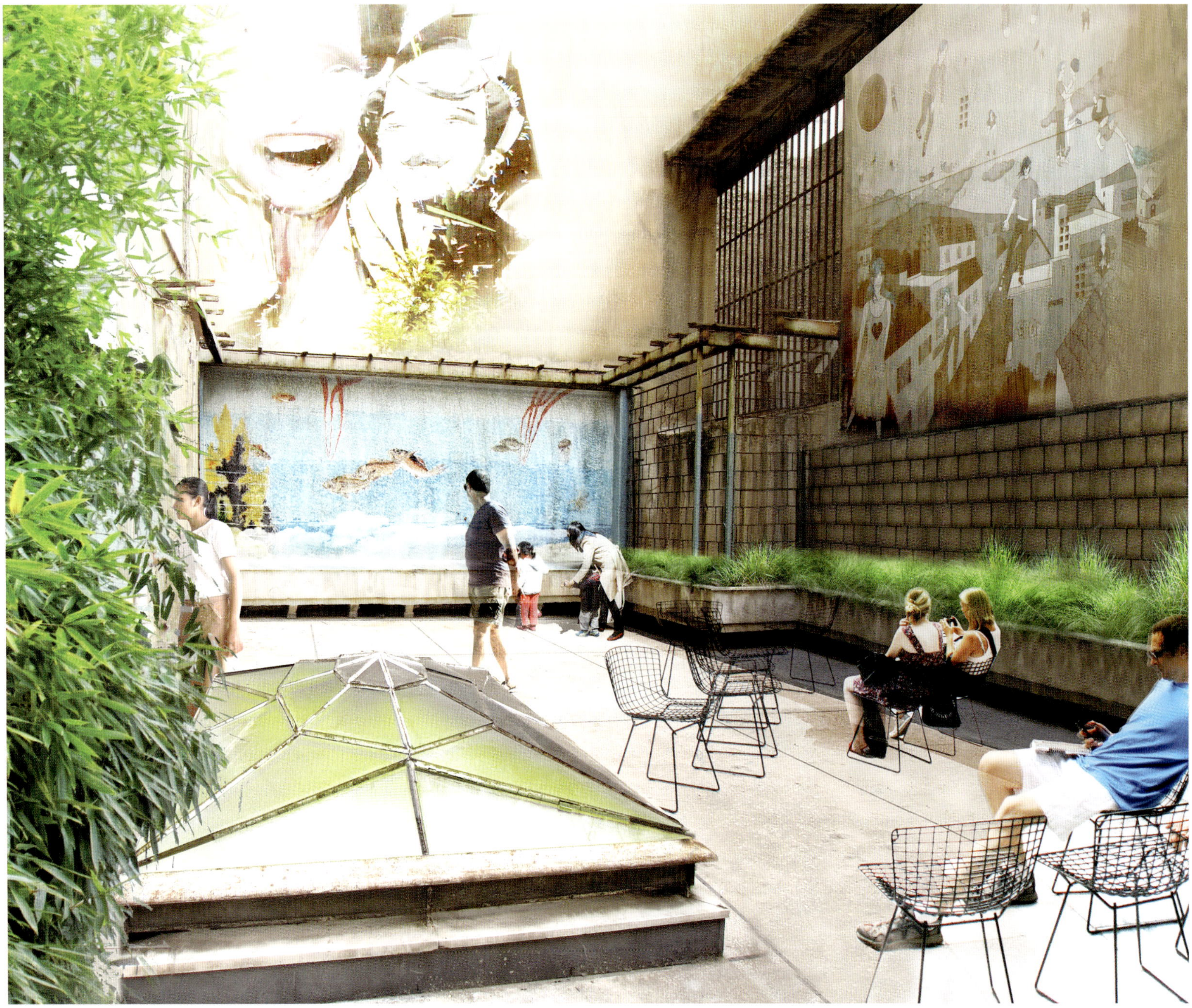

↑ | **Theater of 1,000 rooms,** building strategy
↗ | **Landscaping and water strategy**
→ | **Green and urban spaces**

Rethinking Athens – Towards a New City Center

Athens

Decades of rapid growth in Athens caused infrastructural problems and a social-cultural imbalance, against a background of difficult economic circumstances. OKRA's solution to this was the creation of an integrated proposal to realize a resilient, accessible and vibrant city center, linking this area to adjacent spaces and therefore becoming a catalyst for the whole city. A new network of public and semi-public spaces will provide new and safe routes, creating new places where people can enjoy and ultimately, encourage new developments. The regeneration of the city center will help to improve the environment of the city and activate the area economically.

PROJECT FACTS

Address: Stadiou-Akademias Patision-Amalias, Athens, Greece. **Completion:** 2016. **Urbanists:** Mixst Urbanisme. **Architects:** Studio 75. **Engineers:** Werner Sobek Green Technologies. **Length:** 3,000 m. **Area:** 56,000 m². **Paving:** granite, marble. **Landscaping:** robust green structure, resulting in temperature decrease. **Street furniture:** family of bespoke design street furniture objects. **Context:** transformation of city center.

← | **Green ensemble,** university in urban park
↓ | **Panepistimiou,** bird's-eye view

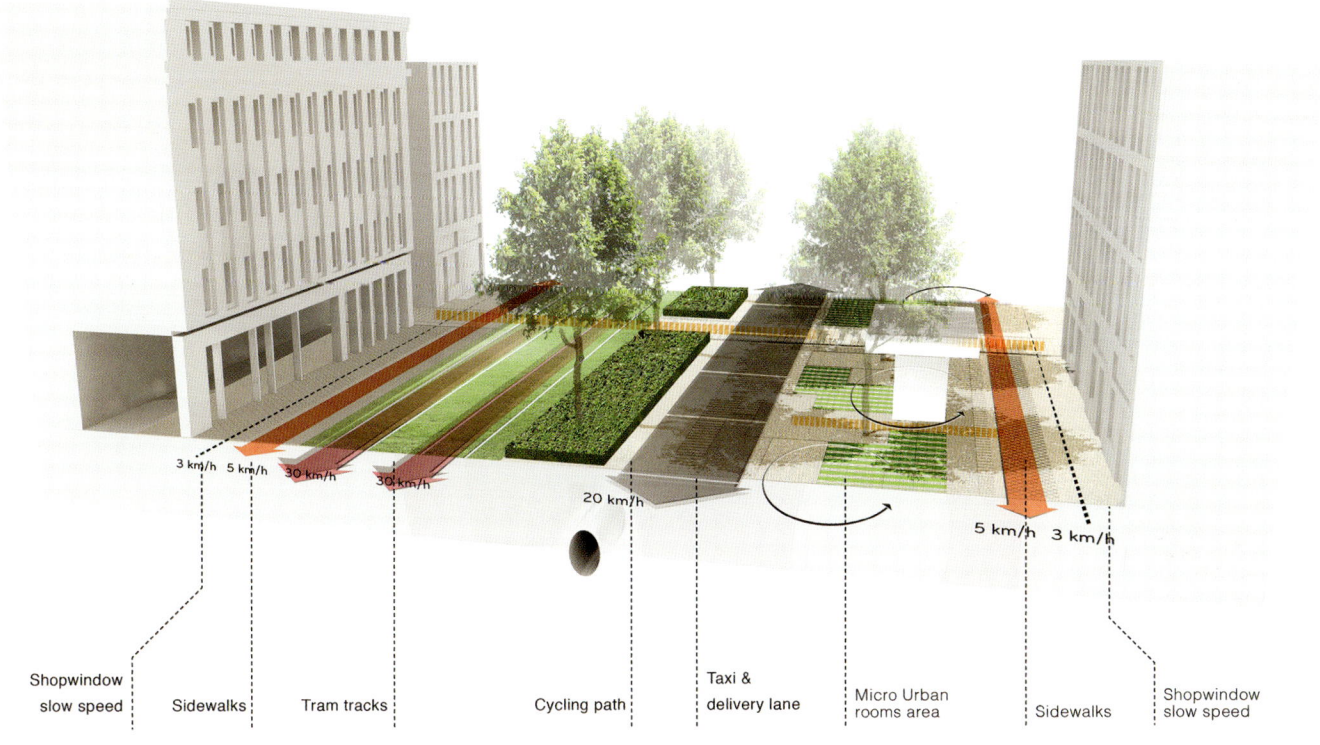

Shopwindow
slow speed

Sidewalks

Tram tracks

Cycling path

Taxi &
delivery lane

Micro Urban
rooms area

Sidewalks

Shopwindow
slow speed

↑ | **Concept program streetscape**
← | **Crossing streets,** new tramline

Mariñas Arquitectos
Asociados

↑ | **Pavilion**
→ | **Wide promenade,** palm trees and integrated seating

Urban Intervention in Port Area

Algeciras

This urban intervention involved the renovation and redesign of the port area in Algeciras, Spain. The historical significance of the site had a profound effect on the design itself. A long row of benches offer space to sit and relax, partially shaded by the overhanging palm trees. Perforated steel elements give the space its unique and unusual character. The pedestrian zone is surrounded by both modern and historical architecture, and attempts to respect the surroundings and present the city with a new modern space to spend time, and enjoy the view.

PROJECT FACTS

Address: Port area, Algeciras, Spain. **Completion:** 2010. **Length:** 260 m. **Area:** 19,500 m². **Estimated visitors:** 500 per day. **Paving:** Spanish granite "Azul platino". **Landscaping:** palm trees. **Street furniture:** integrated stone benches, steel frames. **Context:** renovated pedestrian area near port.

← | **Landscaping follows the line of the street**
↓ | **Steel elements,** remind visitors of the site's industrial heritage

↑ | Use of steel is a continuous theme
↓ | Site plan

↑ | **Entrance situation**
↘ | **Flower bed with rabbit and footprints**

→ | **Two-story mall**
↓ | **Original stamped concrete**, detail

Mitsui Outlet Park

Iruma

This site is located near a vast woodland area and its close proximity to this is reflected by the landscape design. The arabesque motifs are intended to symbolize feelings of love and infinity with nature. The arabesque "Iruma Karakusa" is an important part of the design of the Outlet Park in Iruma. Earthscape was also responsible for the original stamped concrete patterns, the Iruma Karakusa and the fallen leaves. The use of trees and plants was also an integral part of the design, intended to forge a connection between the urban mall complex and the surrounding woodland area.

PROJECT FACTS **Address:** 3169-1 Miyadera, Iruma, Japan. **Completion:** 2008. **Architects:** Laguarda.Low Architects. **Length:** 700 m. **Area:** 86,000 m². **Estimated visitors:** 20,000 per day. **Paving:** concrete. **Landscaping:** trees and small round beds. **Street furniture:** custom-made street furniture references forestry setting. **Context:** shopping mall in wooded landscape.

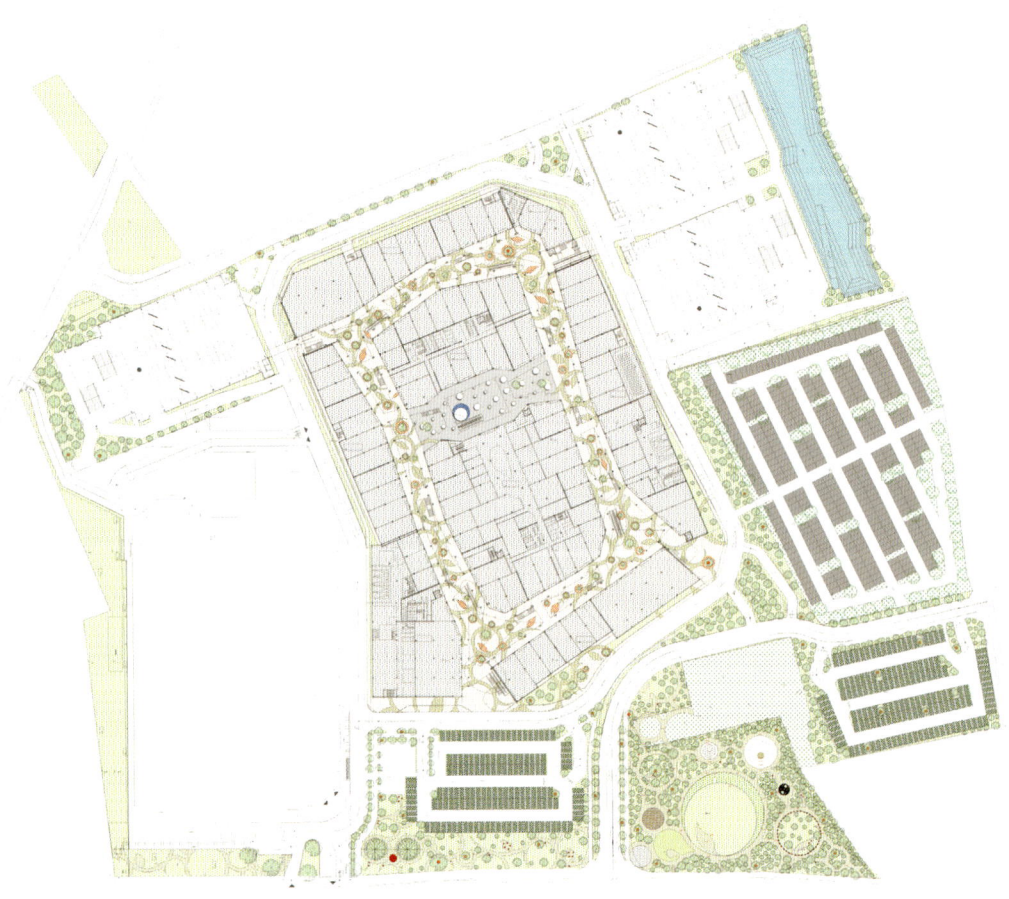

↑ | Fountains
↙ | Layout

↑ | **Paving detail,** original stamped concrete
← | **Site plan**

3:0 Landschaftsarchitektur
Gachowetz Luger Zimmermann

↑ | **Pedestrian route along the ring road**

Ringstraße Sonnenallee

Vienna

This design for the new urban lakeside of Vienna-Aspern is based on the master plan by Tovatt Architects & Planners. The central element of the design is the four-kilometer-long ring road. This curving road doesn't just serve traffic, but also provides an attractive location for pedestrians, cyclists, and people searching for a place just to sit and relax. Hotspots are located along a linear route, each defined by its own unique characteristics. The first section comprises green areas and is for children and young families. The second section is designed to appeal to teenagers, featuring a cycle track, boccie area and arena. The third and final section is for the general public, based on the concept of a picnic blanket.

PROJECT FACTS **Address:** Aspern, Vienna, Austria. **Landscape designers:** Acer Campestre. **Completion:** 2015. **Area:** 17,300 m². **Paving:** concrete pavement. **Context:** ring road and new town center.

↑ | **Ring road,** section of layers
↓ | **Ring road plan**

↓ | **Landscaping along ring road**

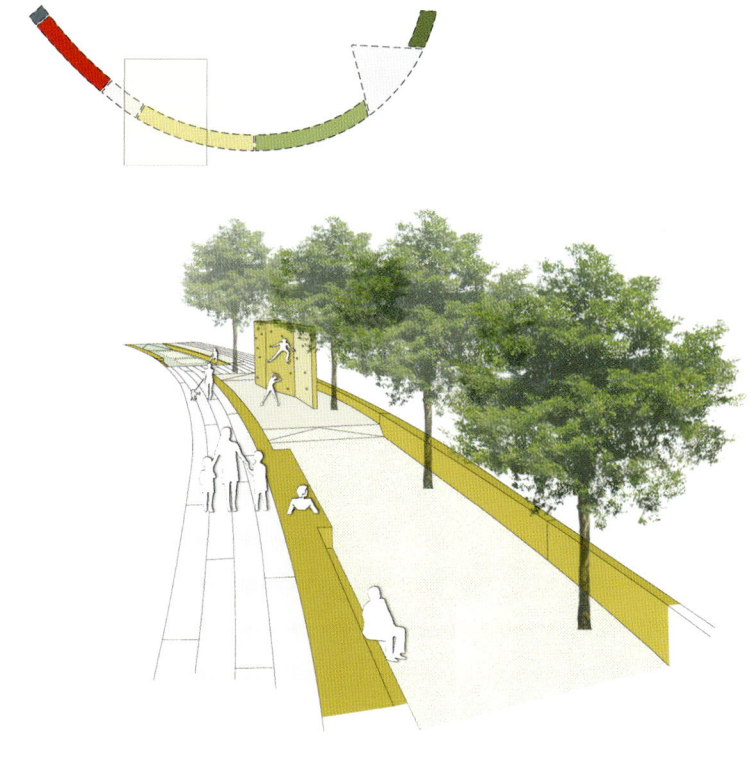

↑ | **Different surfaces come together**

↙ | **Jagged paving stones create a mosaic effect**

↓ | **Paving detail**

The Streets of Herning

Herning

Within this city, the central areas are in the process of being reorganized. The new surfaces can be described as floating fragments that slide and shift among themselves and thus create directions and movements in the upper layers. Flat moorlands characterize the landscape of western Jutland and Herning, located just south of a wash-out plane formed during the last ice age. Water from melting ice has been a great inspiration for the overall surface concept. Ice floes floating down city streets and lanes and takes shape according to the size of the spaces in which the floes assemble.

PROJECT FACTS

Address: city center, Herning, Denmark. **Completion:** ongoing. **Area:** 4,025 m². **Paving:** granite. **Street furniture:** sculpture by Ingvar Cronhammar. **Context:** reorganization and transformation of city center areas.

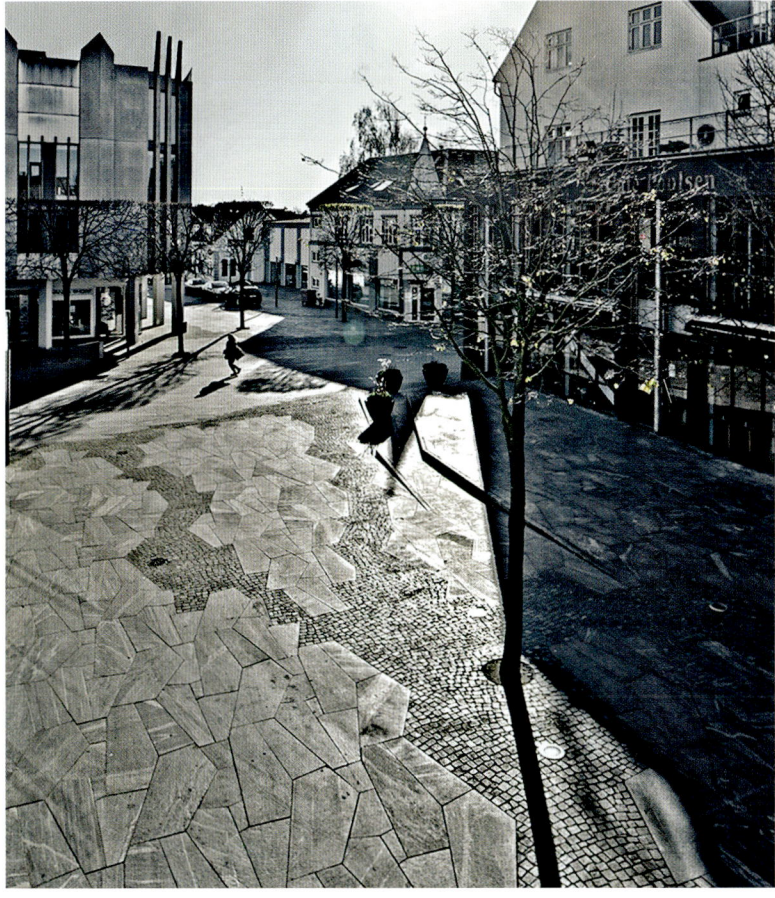

↑ | **Varying paving styles,** reminiscent of cracked ice

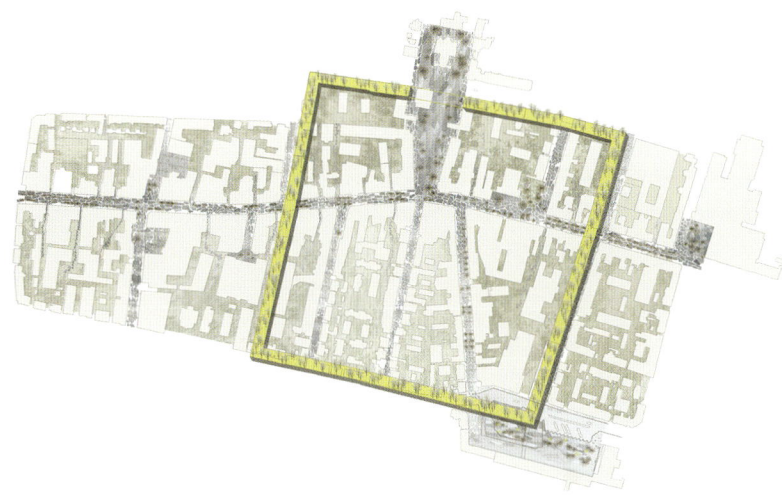

↑ | **Site plan**

↓ | **Cronhammar's sculpture,** bird's-eye view

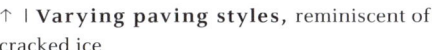

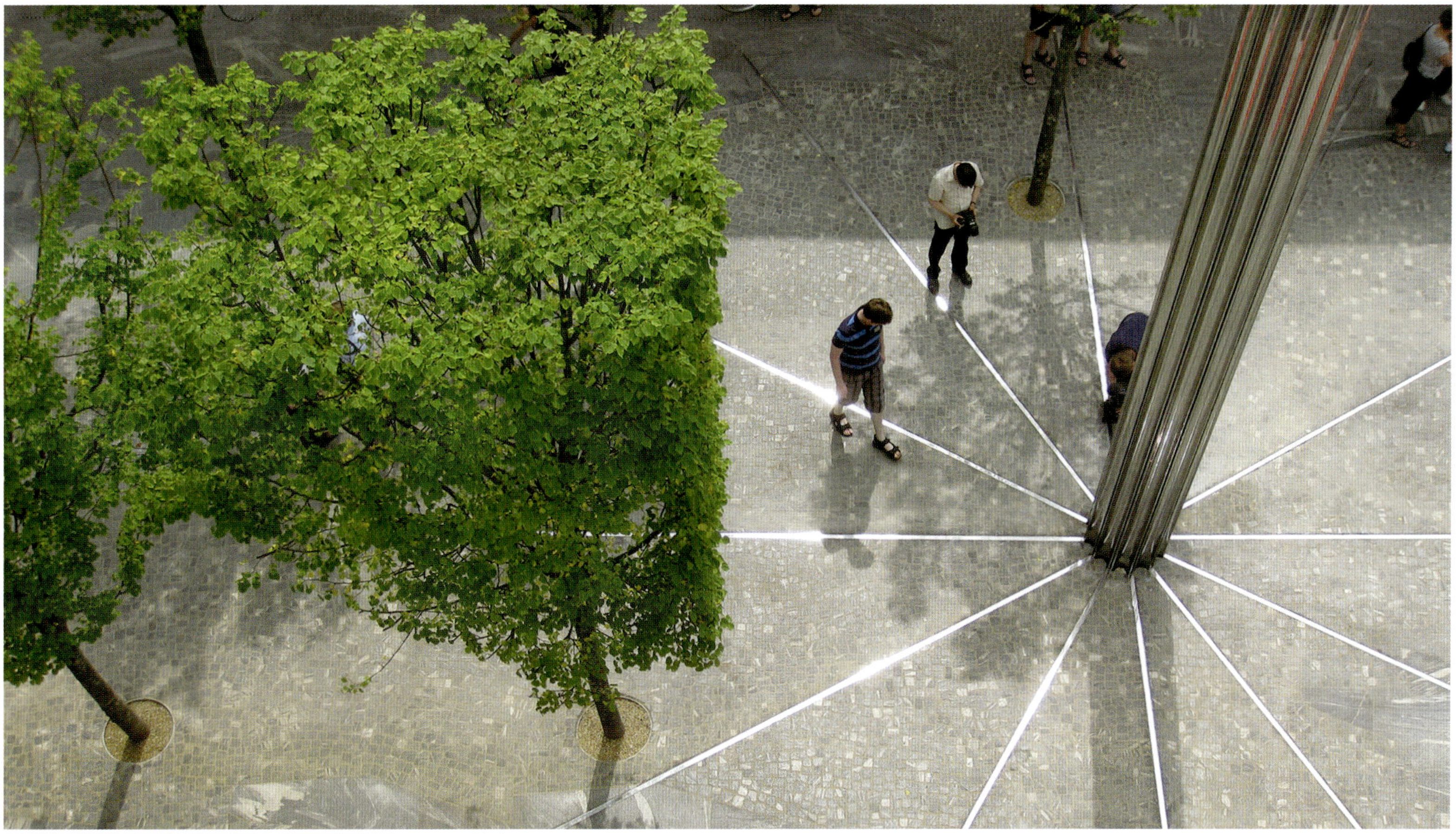

AWP office for territorial
reconfiguration

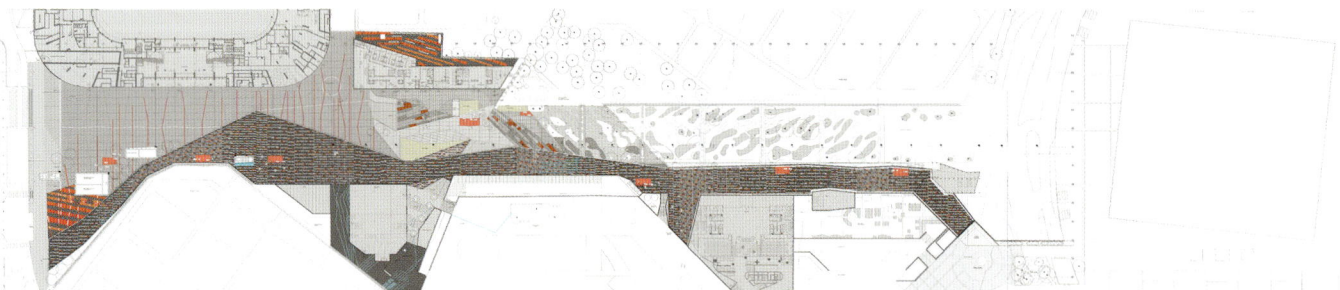

↑ | **Overall view**
↘ | **Site plan**

The Arche's Promenade

La Défense

The principal objective of this design was to create an architectural, urban and landscaping continuity between the newly developed and existing urban spaces. This is one of Grand Paris' major public urban spaces. The main priority is that the site creates a link between two sections of the Seine, asserting its presence at their meeting point. This linking supersedes any private or local issues about the space and perhaps even influences the way these issues are addressed. For this concept, the architects searched for a new intensity of landscape that became a key image for Grand Paris.

PROJECT FACTS

Address: Jardins de l'Arche, La Défense, France. **Completion:** 2016. **Length:** 750 m. **Area:** 71,044 m². **Estimated visitors:** 40,000 per day. **Paving:** trapeze prefab slabs. **Landscaping:** similar species of plants of Gilles Clément gardens (e.g. Salix Alba, Fagur Sylvatica). **Street furniture:** designed by the landscape architects. **Context:** central business district.

↑ | Preservation and integration of existing jetty and its surroundings

↑ | Construction site
↓ | View towards the ramp and its sport buildings

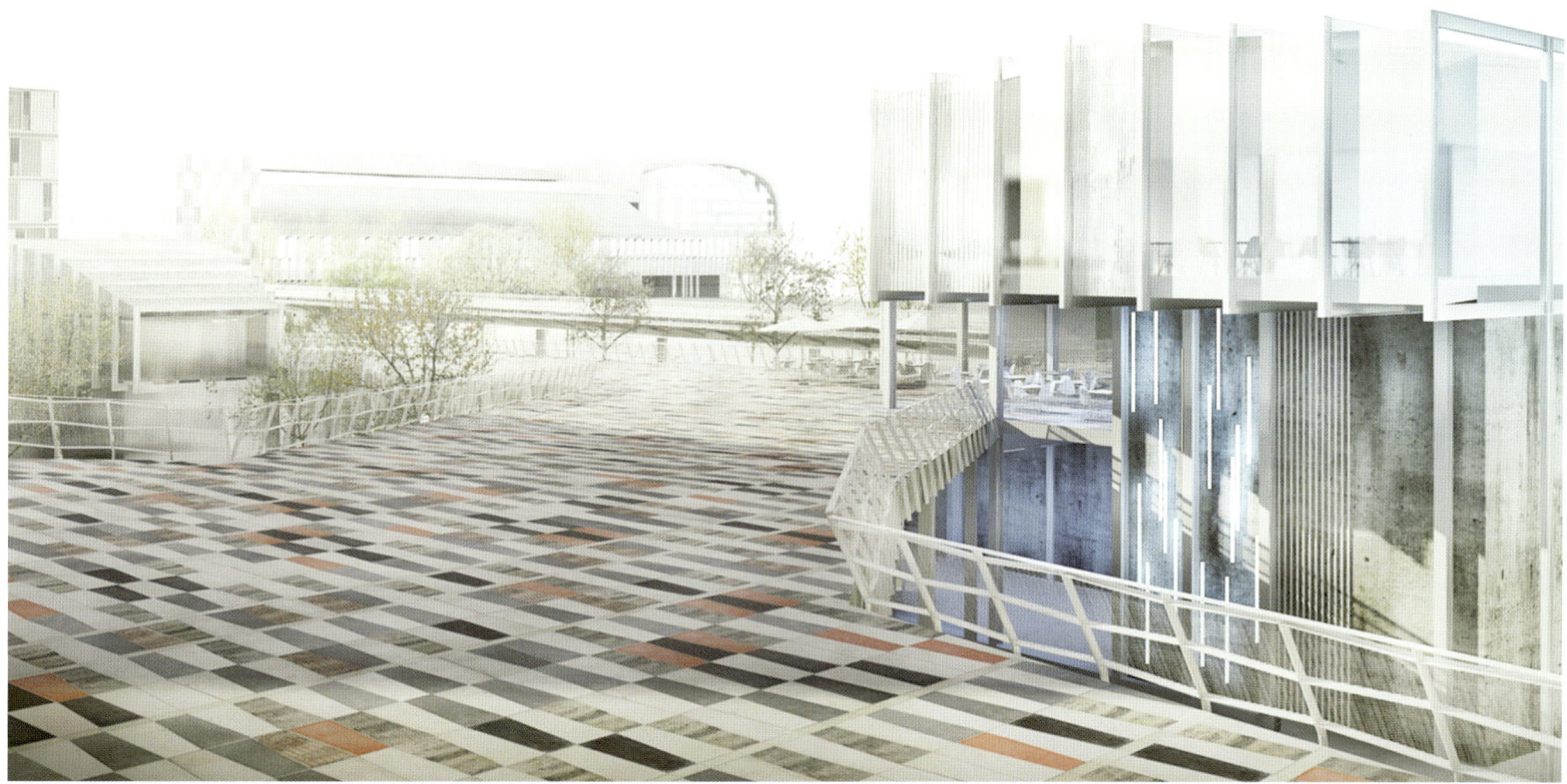

Gustafson Guthrie Nichol

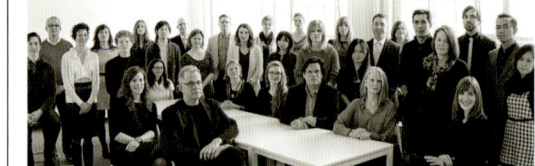

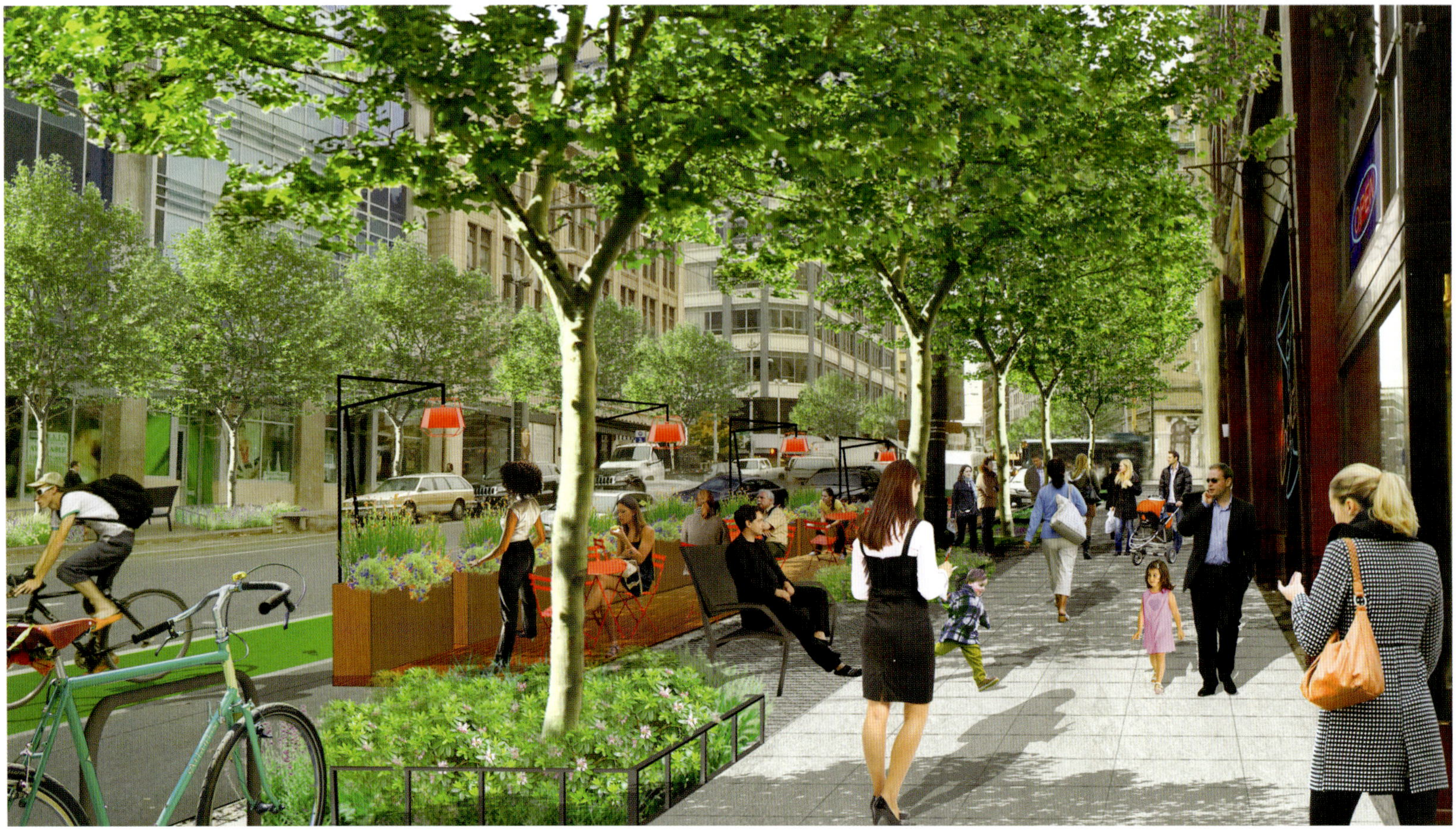

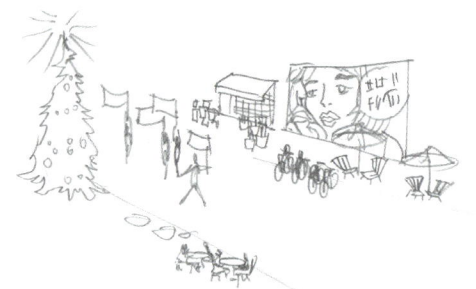

↑ | Avenue
↓ | Concept sketches

Pike-Pine Renaissance

Seattle

The Downtown Seattle Association engaged Gustafson Guthrie Nichol (GGN) to create a detailed plan of phased, opportunistic streetscape improvements to refresh the city's downtown core. GGN's approach to this project was to avoid fancy paving and quickly-dated custom elements that have characterized streetscape improvement districts in the past. Instead, GGN proposed a rediscovery of the simple, urban street standards that create the elegant, timeless base for ever-changing street life and installations. Grand, leafy avenues are made more tailored and green; rugged, irregular hill streets are made more textured and open to views between water and hills. Sidewalks physically continue flush across the traffic lanes of busy avenues to create a comfortable, continuous walking experience for people of all abilities.

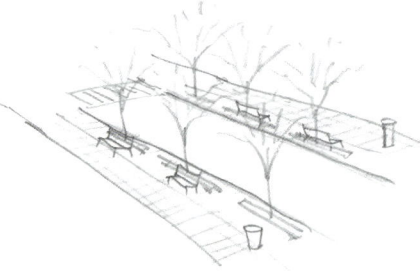

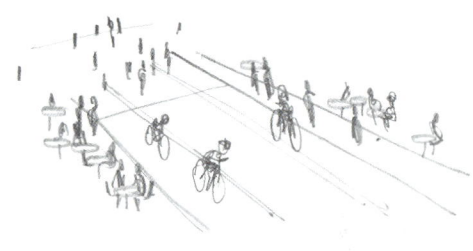

PROJECT FACTS

Address: Seattle, WA, USA. **Urban design consultants:** Framework. **Completion:** 2013. **Area:** 2,630,345 m². **Context:** street improvements in downtown Seattle.

- hierarchy favors the inconvenient
- slows the fast/easy avenues
the hillclimbs are distinct features

↑ | Sketch

PARAMOUNT
macy's

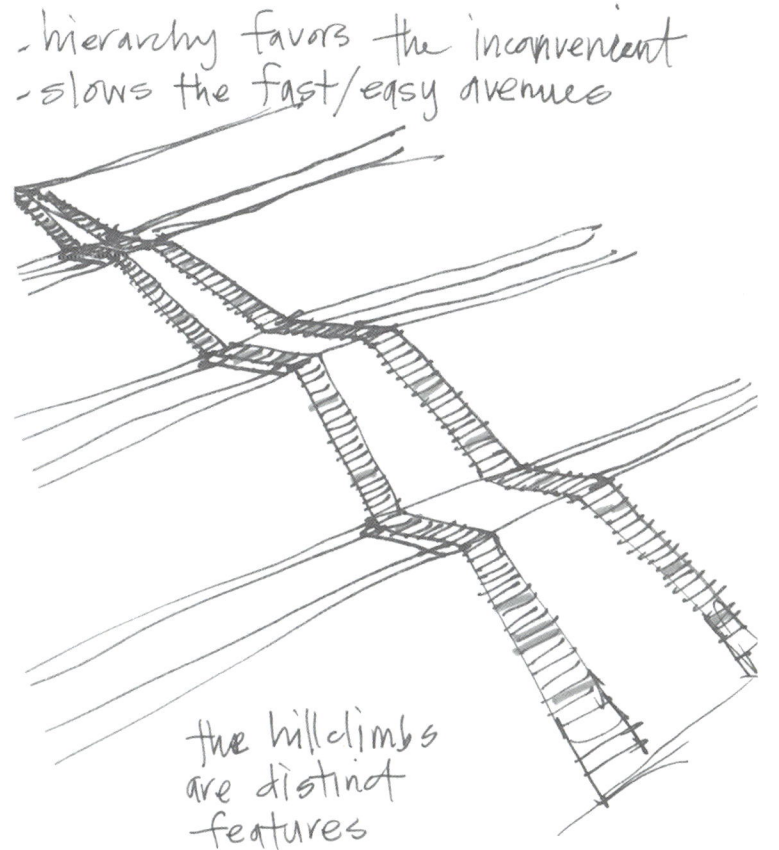

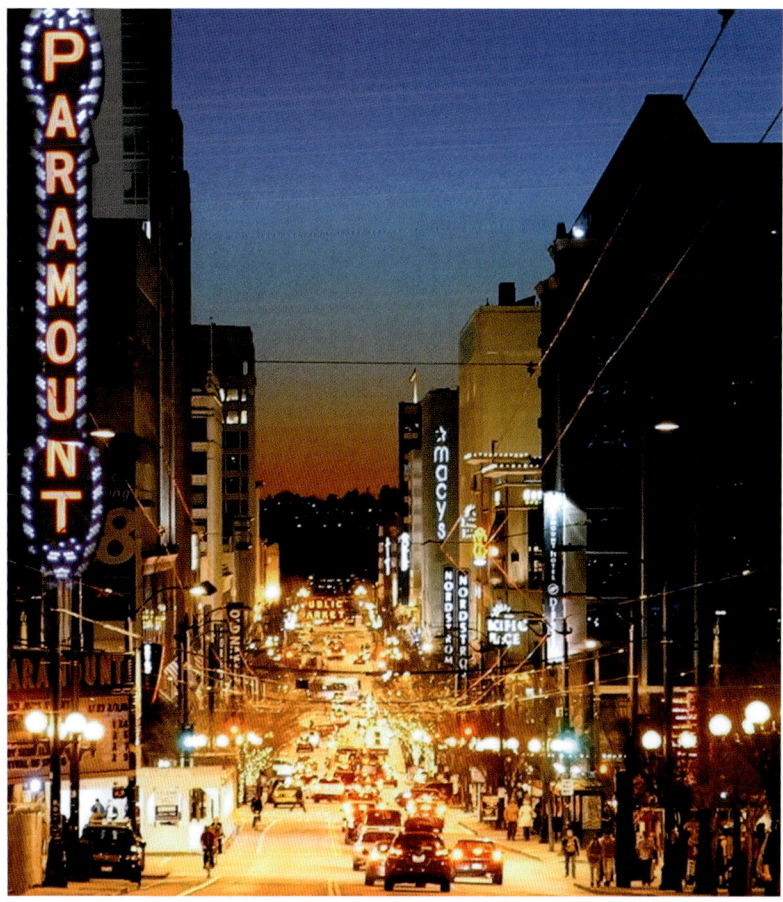

↑ | Hill Street at night

↓ | Downtown Seattle

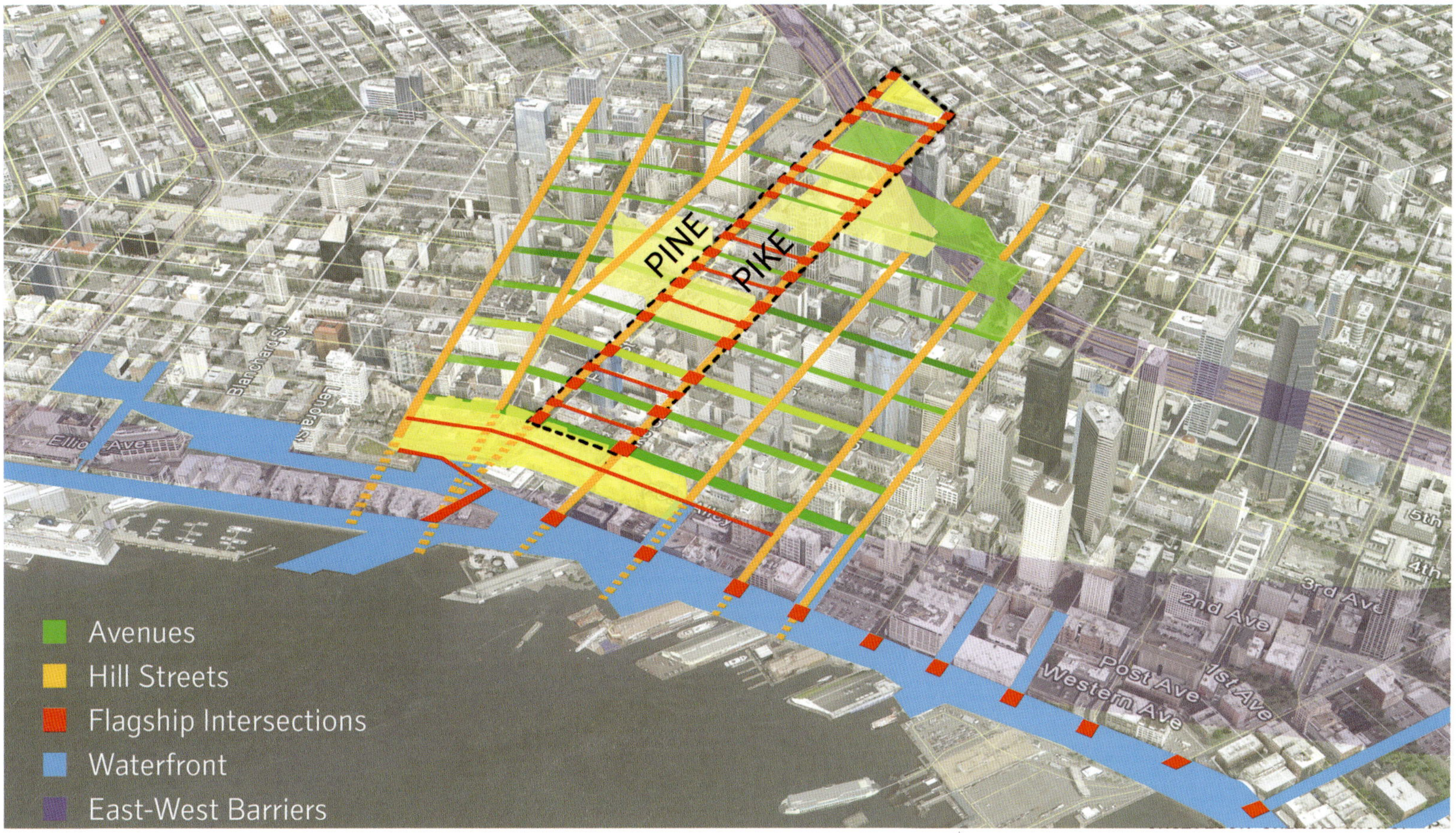

PINE
PIKE
Elliott Ave
Blanchard St
Lenora St
2nd Ave
3rd Ave
4th
5th
Post Ave
Western Ave
1st Ave
Avenues
Hill Streets
Flagship Intersections
Waterfront
East-West Barriers

WES LandschaftsArchitektur

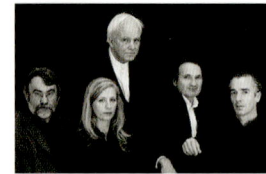

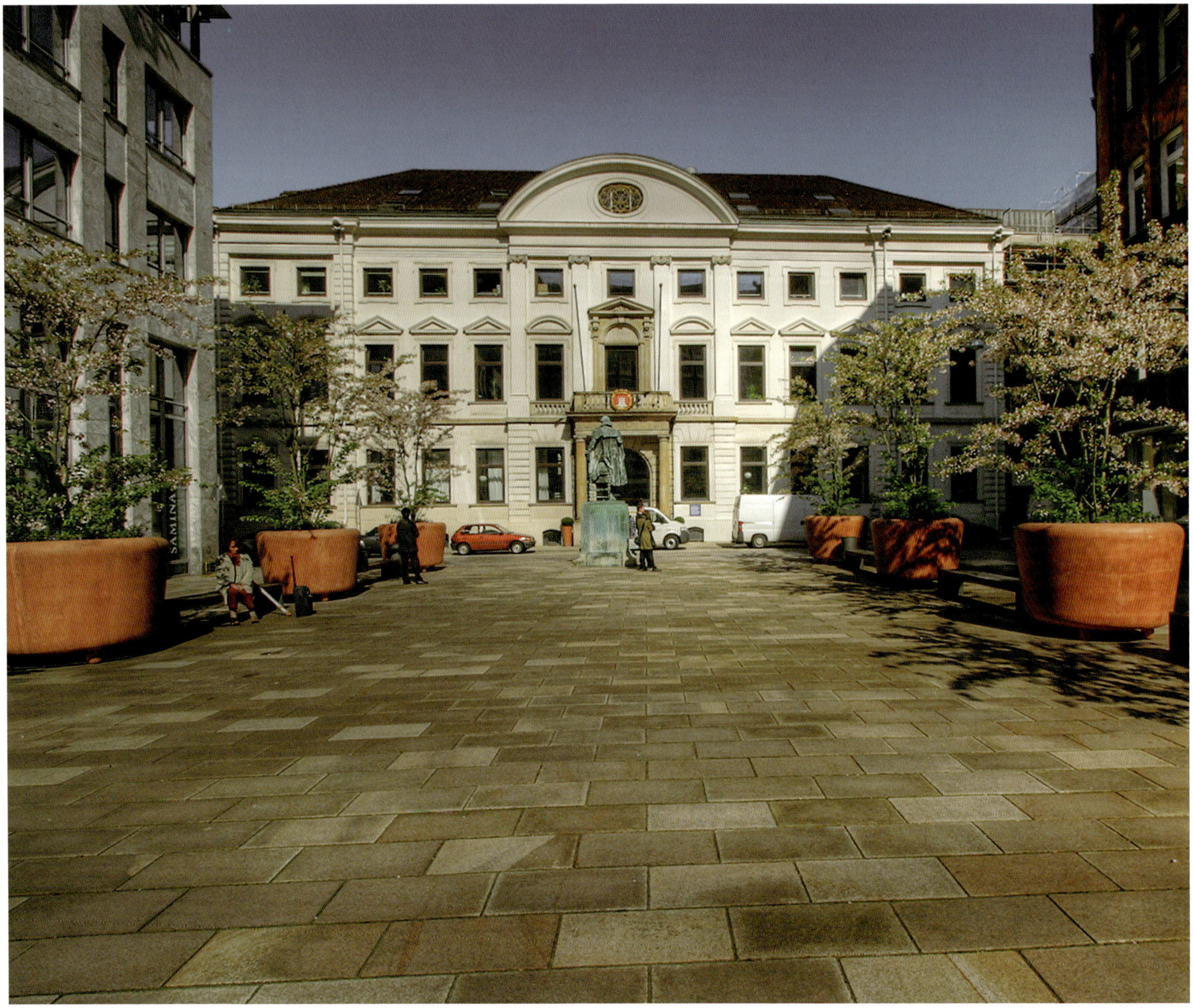

↑ | Landscaping and seating make this an attractive area to sit and relax
↗ | Design combines historic and modern elements
→ | Boulevard in the evening

Neuer Wall

Hamburg

As a result of this redesign, the Neuer Wall street has been transformed back into an exclusive boulevard: it is one of the ten leading luxury shopping streets in Europe and sores as a distinctive and attractive meeting place for all Hamburg residents and visitors to the city. With its newly structured street and enlarged boulevard areas, the Neuer Wall is presented as an elegant urban space with a dash of Hanseatic reserve. At the same time, the light granite on the broad sidewalks and the paved intersections give it a Mediterranean flair. The street is given a touch of green with large plant pots that have been allocated to the individual entrances to the stores. The extensive measures have been made possible by the establishment of Hamburg's first Business Improvement District.

PROJECT FACTS

Address: Neuer Wall, 20354 Hamburg, Germany. **Completion:** 2006. **Area:** 8,500 m². **Paving:** granite. **Landscaping:** pavement and large plant pots. **Street furniture:** benches and lighting. **Context:** redesign of city boulevard.

← | Potted plants line the boulevard
↓ | Site plan

↑ | Design adds Hanseatic flair
← | Shopping street in the evening

Aspect Studios

![Play area photograph]

↑ | **Play area,** interactive water elements
→ | **Playground is a regional attraction**

Darling Quarter

Sydney

Darling Quarter has transformed the public domain of Darling Harbour, one of Australia's most visited destinations. The project includes a new public park, retail outlets, commercial buildings, a children's theater and an innovative children's playground. The playground is the largest and most unique with its interactive water-play facilities and has become a regional attraction. A range of destinational and free place-making initiatives have been implemented including an enlarged park, table tennis tables, moveable public seating and rugs, and a lighting master plan which enriches the night-time experience.

PROJECT FACTS

Address: Darling Harbour, Sydney 2000, Australia. **Completion:** 2011. **Collaborating architects:** FJMT. **Area:** 15,000 m². **Paving:** clay pavers, bricks, concrete unit paving, Australian bluestone, stonevue, interpretive inlay: precast concrete, marine bronze. **Landscaping:** native palm tree planting, sensory planting, rainforest planting, turf. **Street furniture:** custom-designed benches, custom-designed table tennis tables. **Context:** retail area.

← | Site plan
↓ | Sand play area

↑ | Interactive water elements
← | Table tennis area

Michael Singer Studio

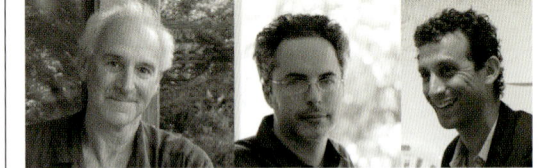

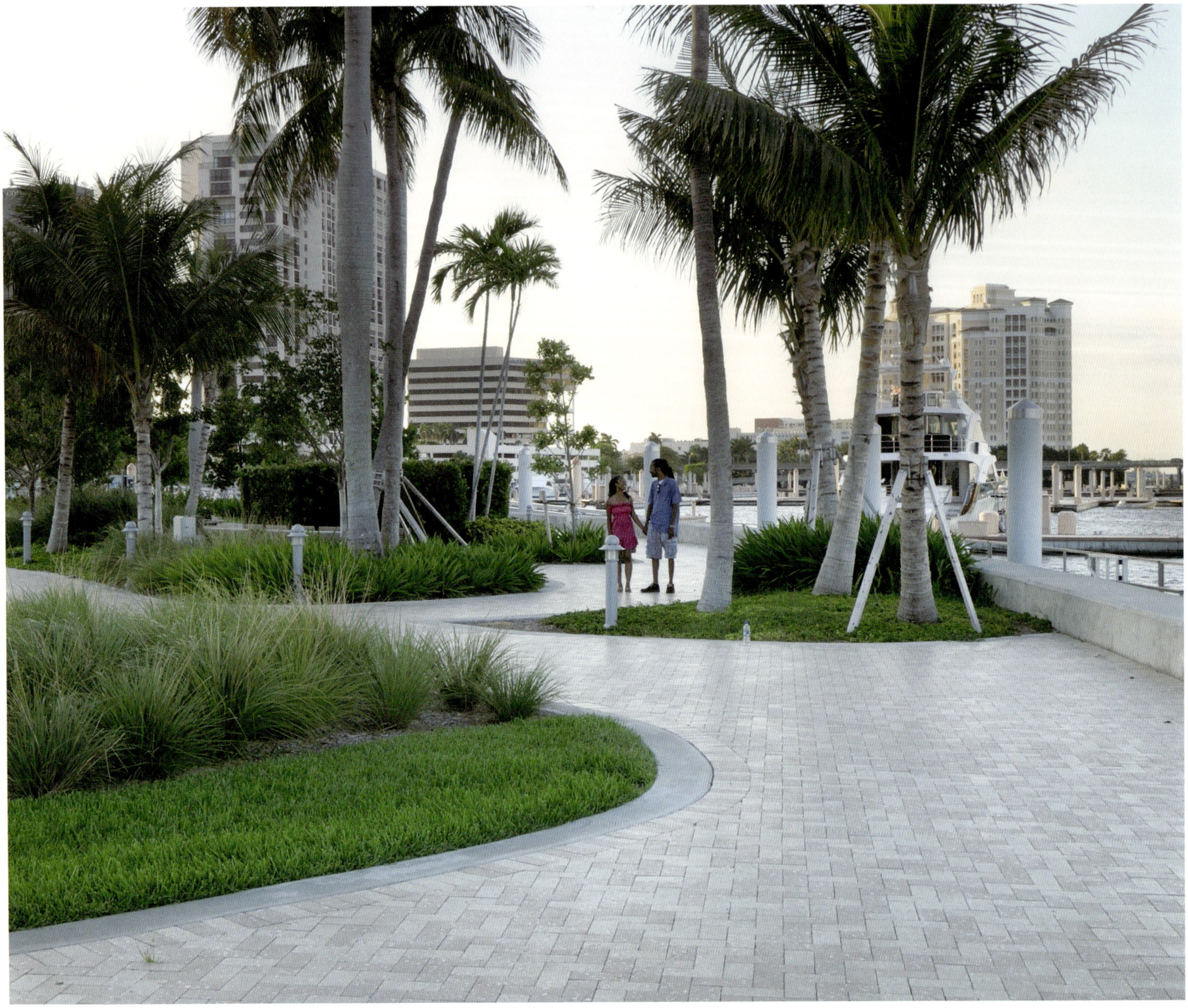

↑ | **Pedestrian route,** stretches along waterfront
→ | **Palms and landscaped islands,** natural appearance

Waterfront Commons

West Palm Beach

Michael Singer Studio was selected as a part of an integrated design team to lead the reimagining and design of the new West Palm Beach Waterfront. The waterfront overlooks the intracoastal waterway with Palm Beach and the Atlantic Ocean to the east. The new civic space revitalizes the city's historic downtown and restores the waterfront's natural beauty. The studio designed the main commons and event spaces, three new floating docks, shaded gardens, two community buildings, a continuous waterfront esplanade, shade trellises, custom benches, seven water elements and an estuarine ecological regeneration area known as the South Cove. The project team included landscape architects Sanchez and Maddux, Studio Sprout and PBC ERM (South Cove).

PROJECT FACTS

Address: West Palm Beach Waterfront, Downtown West Palm Beach, FL, USA. **Landscape architects:** Sanchez and Maddux. **Completion:** 2010. **Area:** 50,000 m². **Paving:** concrete pavers with shell, recycled concrete stone dust. **Landscaping:** mix of tropical and native plants. **Street furniture:** custom concrete and wood benches. **Context:** city waterfront.

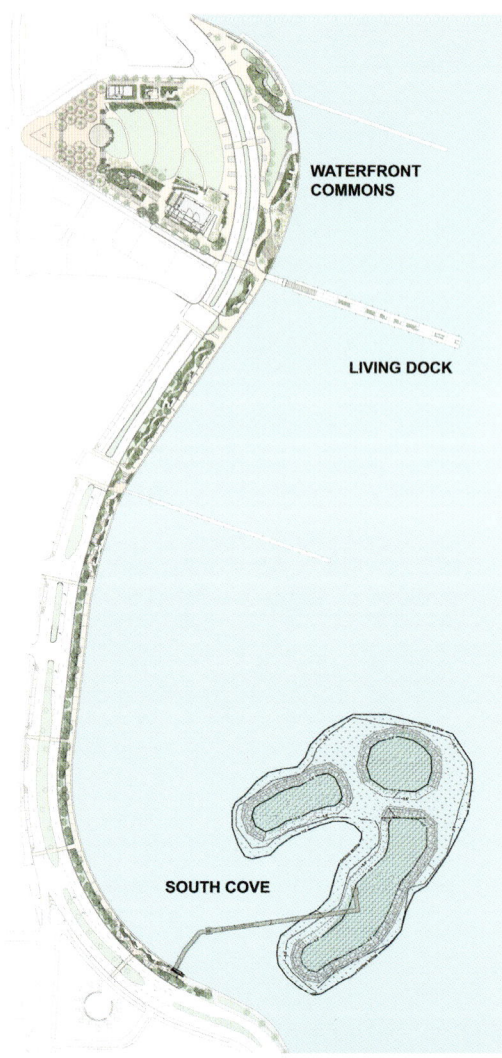

WATERFRONT
COMMONS

LIVING DOCK

SOUTH COVE

← | Site plan
↓ | Aerial view

↑ | **Swing chairs,** looking towards waterfront
← | **Water,** integral part of the design

↑ | **Public space Hoekenrode Plein,** in front
of train station
→ | **Area from above**

Arena Boulevard and Amsterdamse Poort

Amsterdam

The Arena Boulevard and the Amsterdamse Poort have been transformed into Amsterdam's second nightlife district. The current central area has two different faces: the busy, small-scale Amsterdamse Poort shopping center and the spacious but often empty Arena Boulevard. The buildings accentuate the impersonal character of the area. The emphasis in the design for the Arena Boulevard is on breaking up its linear character, and creating a space that is not only pleasant for a group of ten people, but also for a crowd of fifty thousand. Long benches of natural stone and wood mark the transition between places for movement and places to pause. A three-dimensional lighting web of cables and spotlights creates a dramatic 'starry sky' above the Arena Boulevard.

PROJECT FACTS

Address: Hoekenrodeplein, 1102 BR Amsterdam, The Netherlands. **Completion:** 2014. **Lighting design:** Karres en Brands. **Area:** 105,000 m². **Estimated visitors:** 5,000–80,000 per day. **Paving:** ceramic bricks in ten colors. **Street furniture:** benches integrated into landscaping concept. **Context:** city center.

ZEEMAN
ici TIEMESSEN

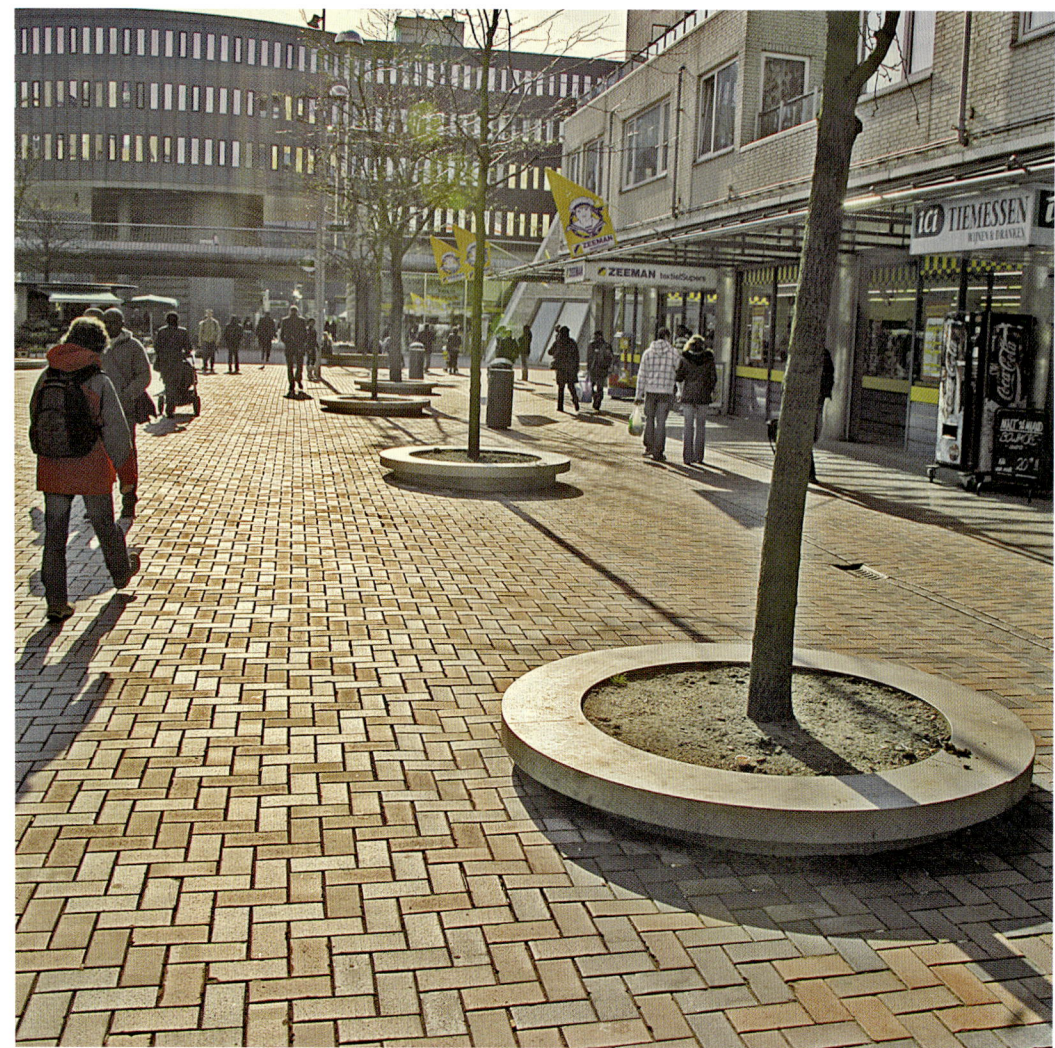

← | **Shopping area Amsterdamse Poort**
↓ | **Site plan**

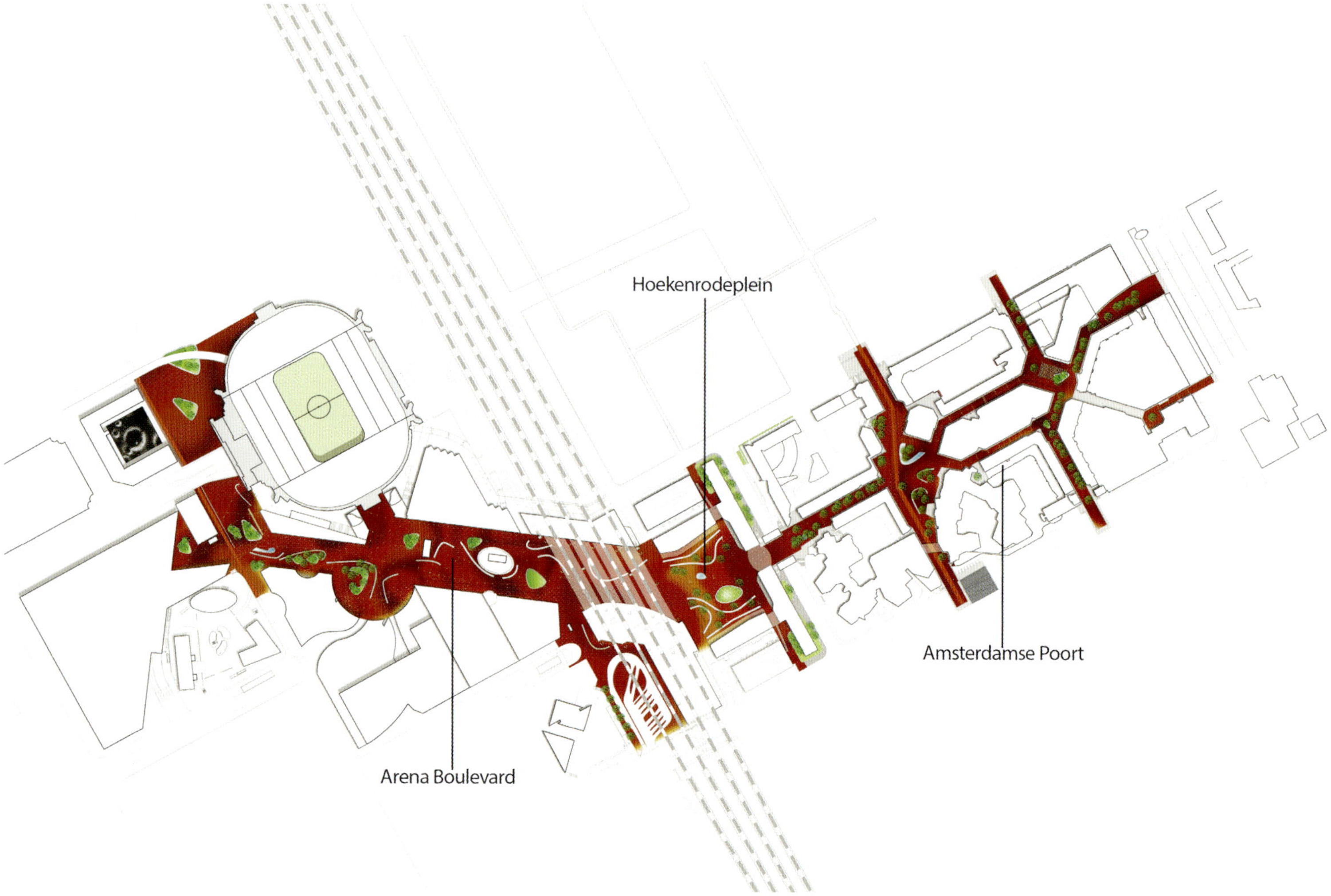
Hoekenrodeplein
Amsterdamse Poort
Arena Boulevard

↑ | **Arena Boulevard,** seating is an integral part of the design
↓ | **Paving**

Atelier Jacqueline Osty &
Associés

↑ | **Promenade,** integrated seating
↗ | **Beachfront**
→ | **Multifunctional area,** pedestrians and
cyclists share space

Seafront Promenade

Les Sables d'Olonne

In most coastal towns, the waterfront is a major axis of movement. This project aims to
reclaim the waterfront in Les Sables d'Olonne by reducing the amount of traffic, reveal-
ing the diversity of the urban space and establishing a more fluid relationship between
the beach and the city. A main focus of the design involved the development of a broad
promenade running the entire length of the beach. This allows pedestrians to move easily
along the bay from the port.

PROJECT FACTS

Address: Les Sables d'Olonne, France. **Completion:** 2013. **Architects:** Lancereau et Meyniel. **Lighting designers:** Concepto. **Area:** 45,900 m². **Paving:** pattern of different types of granite. **Street furniture:** long wooden benches integrated into promenade. **Context:** city seafront.

← | **Shaded seating,** view of sea
↓ | **Promenade,** fuses beach and city together

↑ | Wealth of places to sit and relax
↓ | Site plan

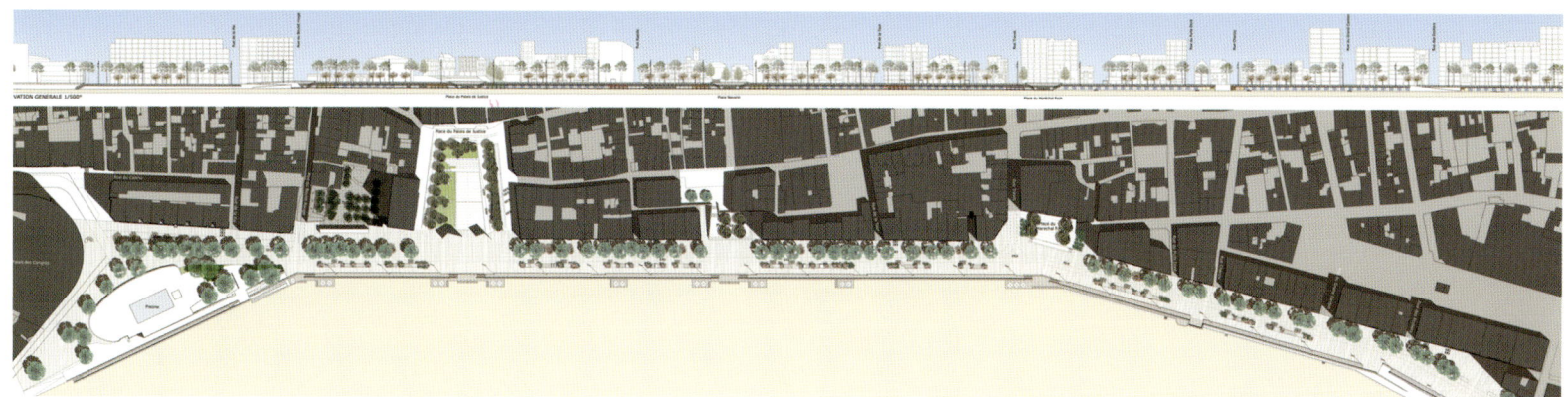

L35 Architects

↑ | **Pedestrian area**
↘ | **Street furniture**, detail

→ | **Lake with sculptural seating elements**
↓ | **Pavement**, detail

Puerto Venecia

Zaragoza

Puerto Venecia is a large-scale project to the south of Zaragoza in a newly developed area well-connected to the city center. Designed along a central axis with water as its main attraction, it features a canal and navigable manmade lake. These are complemented by green spaces and pedestrian and recreational areas, creating a 25,000-square-meter urban park which is surrounded by shops, restaurants, entertainment and sports facilities. High-quality materials and state-of-the-art urban furniture designs have been used throughout to create comfortable and welcoming spaces for visitors to enjoy leisure and culture. A wide range of attractive plants, ornamental fountains and sculptures help to create a quality urban environment for this project based on environmentally friendly principles.

PROJECT FACTS

Address: Travesía Jardines Reales 7, 50021 Zaragoza, Spain. **Completion:** 2012. **Landscape designer:** Mike Smith. **Lighting designer:** Theo Kondos. **Length:** 1,000 m. **Area:** 25,000 m². **Estimated visitors:** 46,500 per day. **Paving:** granite, concrete. **Street furniture:** sculpture by Arturo Berned, Santiago Gimeno. **Context:** urban park in shopping area.

← | **Pedestrian area,** between lake and canal
↓ | **Pedestrian area**

↑ | **Play area**
↓ | **Site plan**

Atelier Jacqueline Osty &
Associés

↑ | **Garden of the Dells,** new mall on the left
↗ | **Hoche Garden**
→ | **Honor Courtyard,** fountain

Parc de la Caserne de Bonne

Grenoble

Located on the grounds of a former military barracks, the park de Bonne is a structuring element in Grenoble's new eco-district. It comprises three separate entities: the Hoche Garden, the Honor Courtyard, and the Garden of the Dells. The park echoes the past history of the site while asserting a contemporary composition. As a reminder of the site's military past, the Honor Courtyard has a rigorous square form. The three gardens form one unique park, clearly noticeable in its continuity thanks to the main circulation axis from east to west. The park should be seen as an axis sequenced by gardens. It is interrupted by multiple pathways that ensure the permeability of the park area.

PROJECT FACTS

Address: 38000 Grenoble, France. **Completion:** 2010. **Lighting designer:** Concepto. **Length:** 450 m. **Area:** 45,000 m². **Paving:** concrete. **Landscaping:** native plants. **Street furniture:** made-to-measure benches. **Context:** part of new eco-district on the site of former military barracks.

← | **Detail,** alternation of vegetation
↓ | **General plan,** paving three gardens

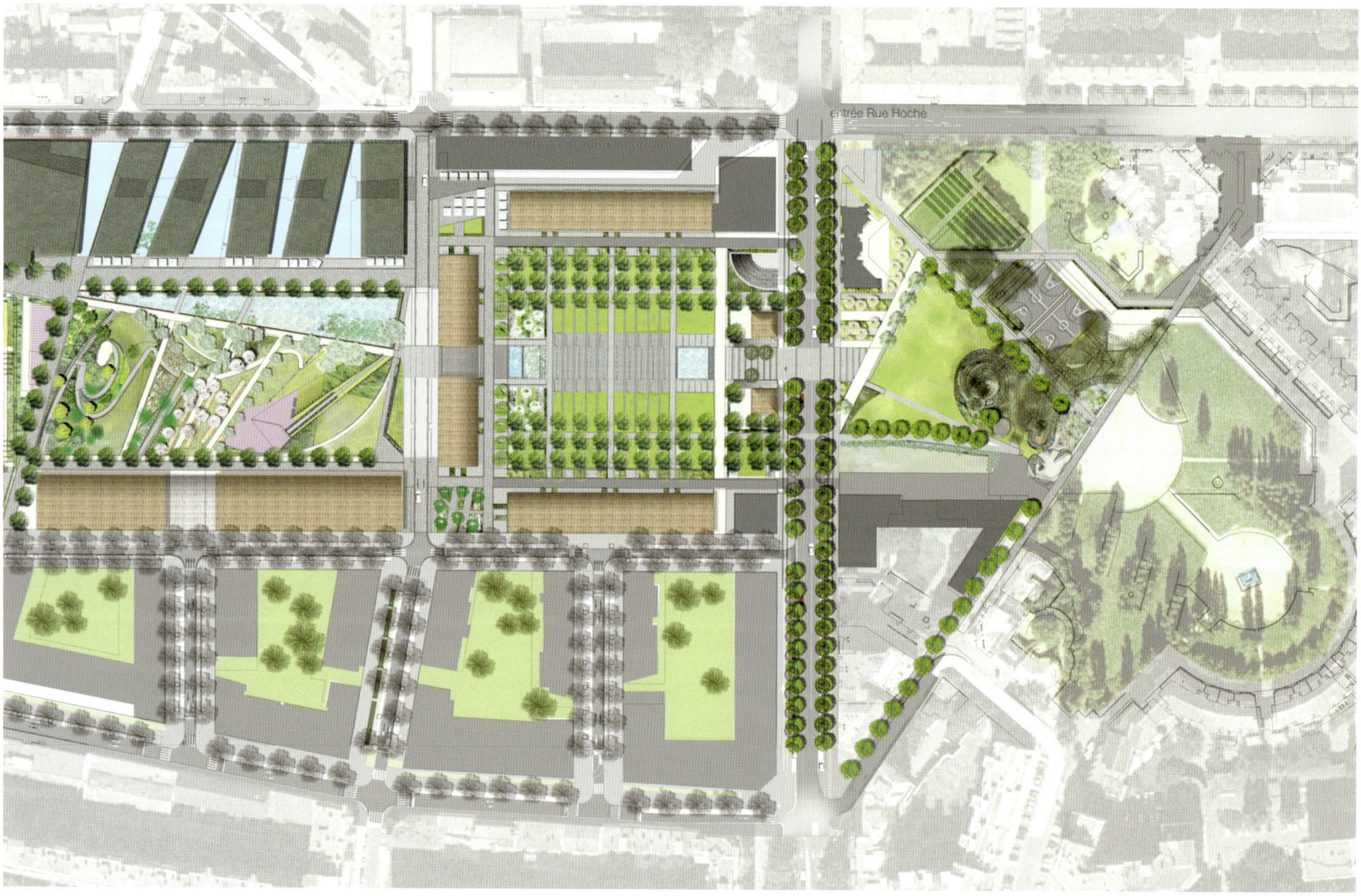

↑ | **Honor Courtyard,** relaxation and play area
← | **Detail,** reuse of stones from military building

Tabanlioğlu Architects

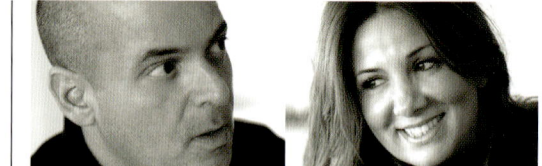

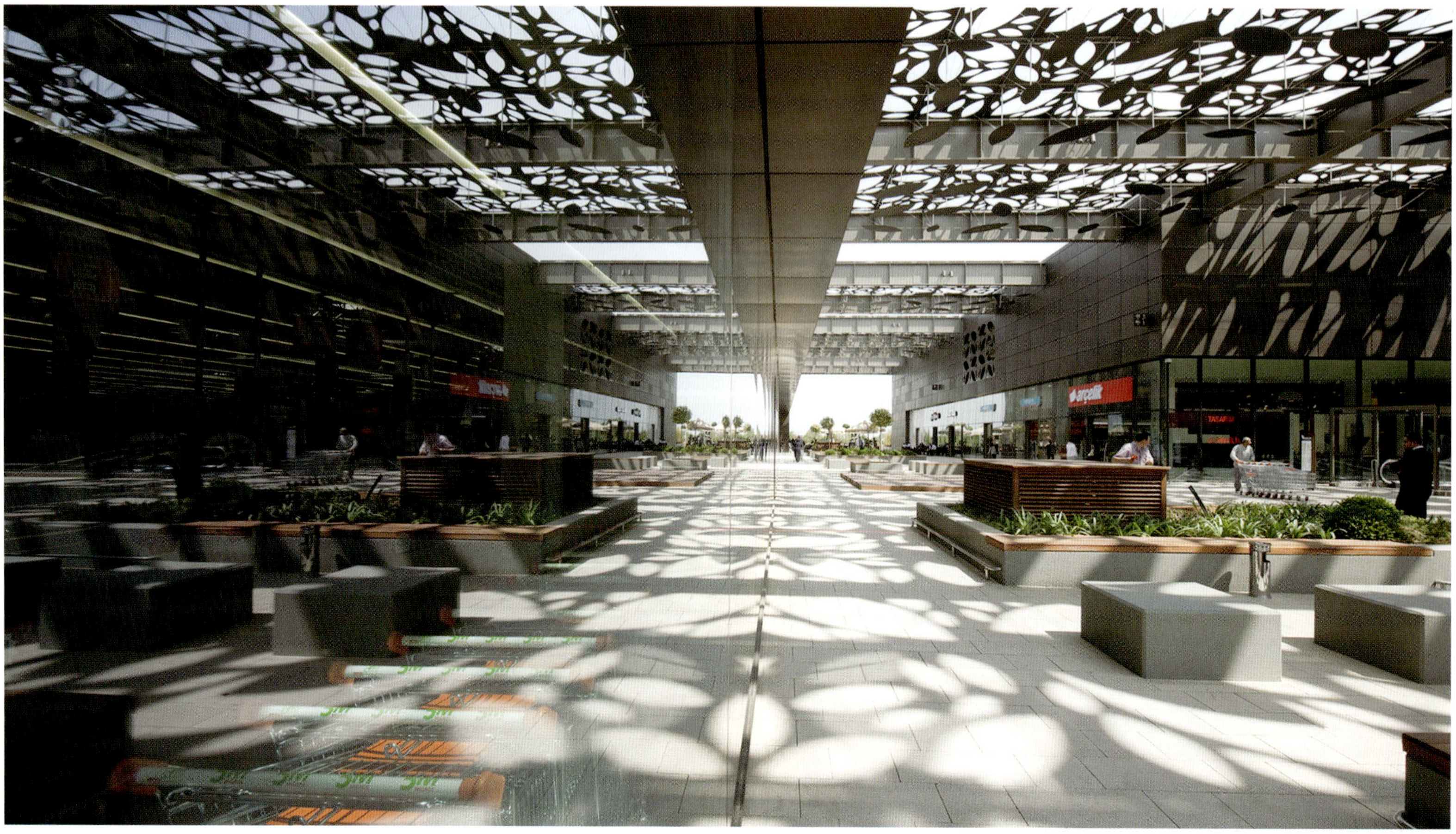

↑ | Play of light and shadow

↙ | Glazing

↓ | Restaurant seating

Asmaçati

Izmir

Asmaçati shopping and meeting point is a semi-open shopping facility in Balçova, inspired by the climate and Aegean tradition of spending time outdoors. Asmaçati provides room for social and cultural activities such as small concerts and exhibitions. Within the framework of sustainability, public space, shopping and street life complement each other in an environment where dissimilar identities get together independent of age, gender and socio-economic differences. In addition to the anchoring hypermarket store that supplies the needs for a wide range consumer goods, 18 shops and restaurants were designed in a boutique concept. The common ground of the visitor profile is the friendly conversations of the Aegean Coast.

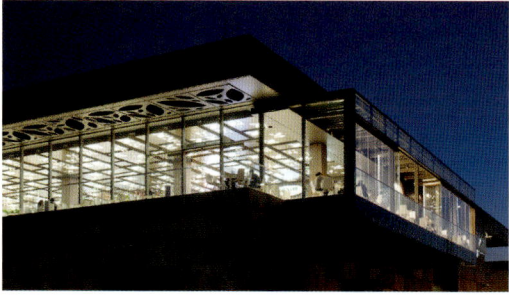

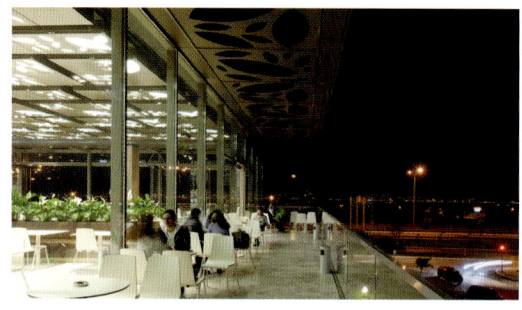

PROJECT FACTS
Address: Mithat Paşa 1460, Izmir, Turkey. **Completion:** 2011. **Interior designers:** Tabanlıoğlu Architects.
Area: 22,763 m². **Context:** shopping center in urban context.

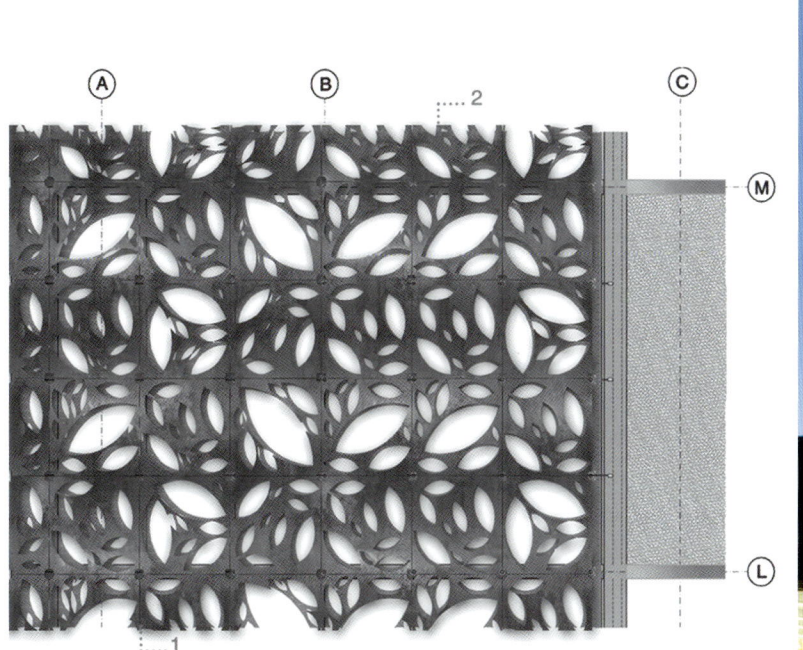

↑ | Patterned roof detail

↑ | Canopy roof with pedestrian area below
↓ | Building overview

M&N Environmental Planning Institute

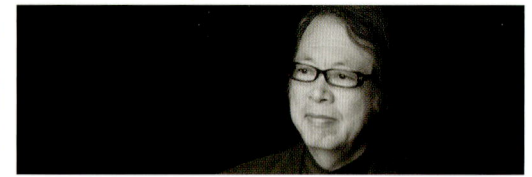

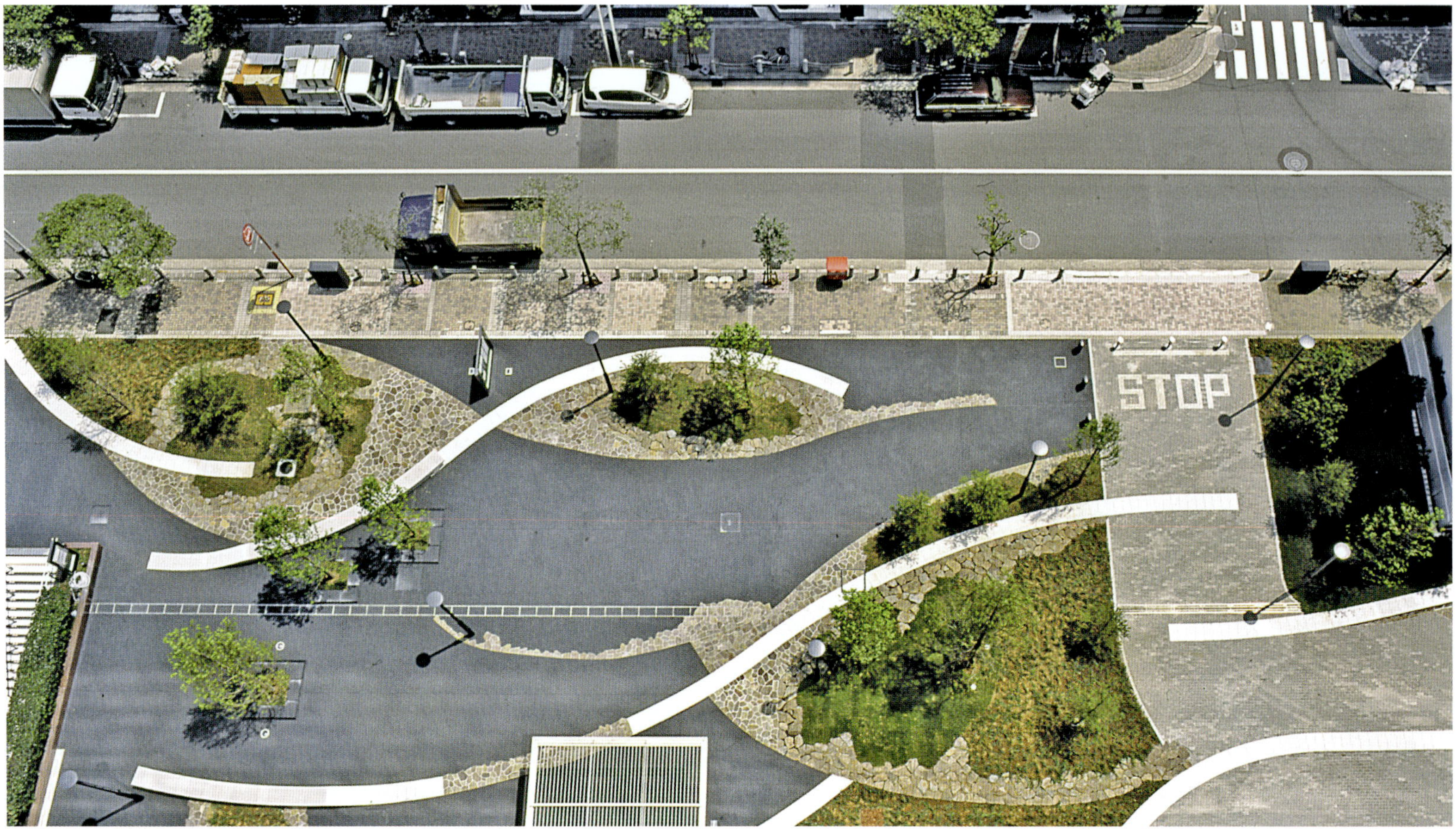

↑ | **Overhead view,** benches and bedrocks form 'small islands in the waves'

↙ | **Plan**

↓ | **Sketch**

Shiba Park Building

Tokyo

As a response to the competition brief, these landscape architects decided to turn this "space" into a "place" – a location where people would want to spend time, where the abundant nature revives the urban environment. The design reflects the ancient Japanese spatial conceptions of "Ke" and "Ma". "Ke" is the idea that there is some kind of energy force around existing objects, while "Ma" can be defined as voids between objects. Due to the understanding of the shift from materialism to spiritualism, this area was reborn as the lively place, spontaneously supporting a range of different activities.

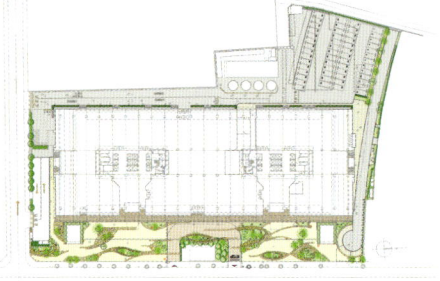

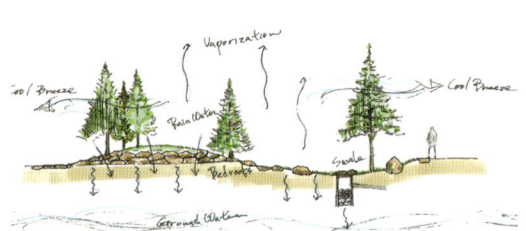

PROJECT FACTS

Address: 2-4-1 Shiba-Koen Minato-ku, Tokyo 105-0011, Japan. **Completion:** 2007. **Length:** 160 m. **Area:** 16,302 m². **Estimated visitors:** 10,000 per day. **Paving:** porous concrete, teppei stone, granite. **Landscaping:** trees (Metasequoia, spruce etc.), groundcover (Juniperus chinensis etc.), stones. **Street furniture:** stone benches designed by Koshi Ohashi. **Context:** pedestrian park area between office building and street.

↑ | **Stone bench,** Metasequoia trees provide shade

↑ | **Bird's-eye view,** project site between building and street
↓ | **People resting,** benches in the shade

Aspect Studios

![Space to play and exercise]

↑ | **Space to play and exercise**
→ | **Bird's-eye view**, cafés, shops and artwork

The Goods Line

Sydney

The Goods Line is a unique new public space designed and led by Aspect Studios, with Chrofi, for the Sydney Harbor Foreshore Authority. Located in Ultimo in inner Sydney, the site is a former historic rail line, which is now bound by a unique concentration of Sydney's key cultural, educational, and media institutions. The 500-meter long elevated civic space will not only transform this industrial relic on the city's western fringe into an innovative example of urban green space for the Sydney suburb of Ultimo, but it will also have a transformative social and environmental role in the precinct by creating a new platform for public engagement and will feature a series of elevated green spaces or platforms.

PROJECT FACTS

Address: Ultimo, Sydney 2007, Australia. **Completion:** ongoing. **Design partner:** Chrofi. **Length:** 500 m. **Area:** 12,980 m². **Paving:** precast concrete. **Landscaping:** native wildflowers, grasses and turf. **Street furniture:** custom steel, precast and timber furniture. **Context:** high density urban environment.

← | Communal area
↓ | Sections

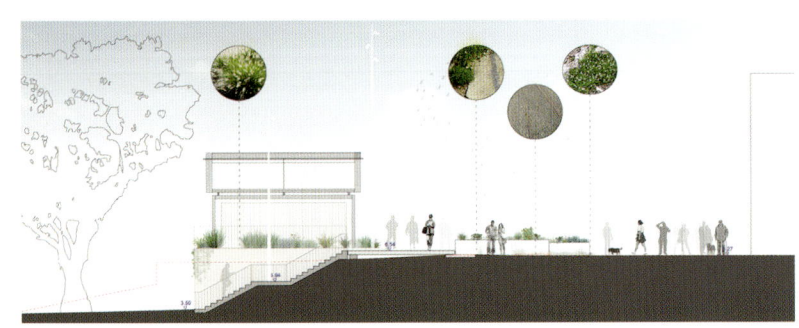

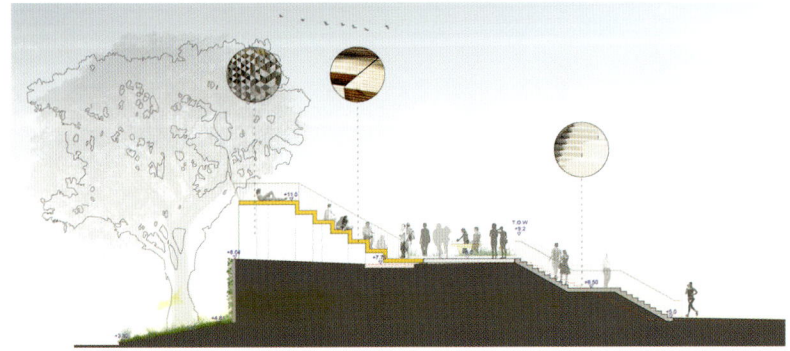

← | Site plan
↓ | View from above

↑ | **Arial view of the promenade,** by night
↓ | **Translucent polycarbonate benches**
↘ | **Retail space within the park**

↗ | **Densely planted raised terraces,** lower heat island effect
→ | **Rest area,** porous paving for stormwater filtration

Gubei Pedestrian Promenade

Shanghai

The creation of a three-city-block pedestrian promenade in Shanghai, China, was a pedestrian oasis in the midst of 20-story high-rise residential towers. Bringing life to the residential area by activating the street level, this project interweaves retail, open space, residential life, and human occupation. Centrally located in the densely populated Gubei District, the promenade provides a framework for the interconnected social ecologies in these progressive neighborhoods. With little differentiation between indoor and outdoor spaces, the landscape design creates a fluid connection between building and open space, putting into form the notion of an "outdoor living room" for public and civic engagement.

PROJECT FACTS

Address: between Gubei Road, Hongbadshi Road, Guyang Road, Shanghai, China. **Completion:** 2009. **Length:** 700 m. **Area:** 46,000 m². **Paving:** SWA designed porous paving and custom paving for graphic patterns. **Street furniture:** custom-designed by SWA. **Context:** pedestrian promenade in urban context.

AXONOMETRIC VIEW OF UNIT

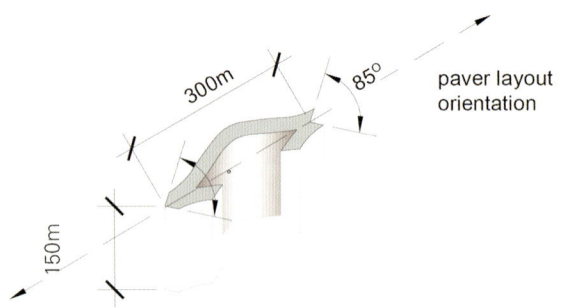

300m

85°

150m

paver layout
orientation

DETAIL SECTION

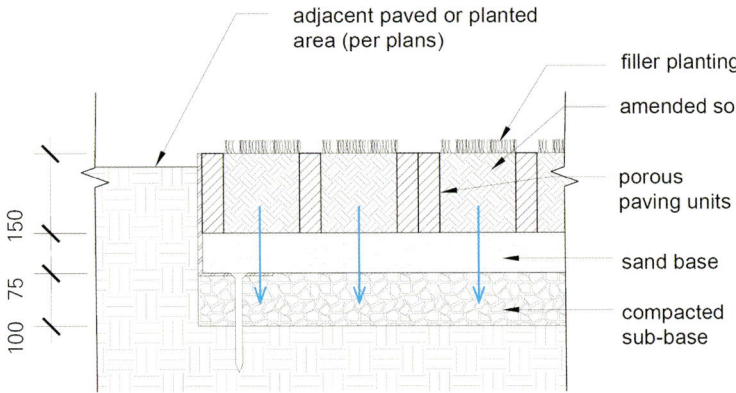

adjacent paved or planted
area (per plans)

filler planting

amended soil

porous
paving units

sand base

compacted
sub-base

150

75

100

← | **Stormwater filtration system**
↓ | **Sculptural water feature,** green glass
tiles and black stone

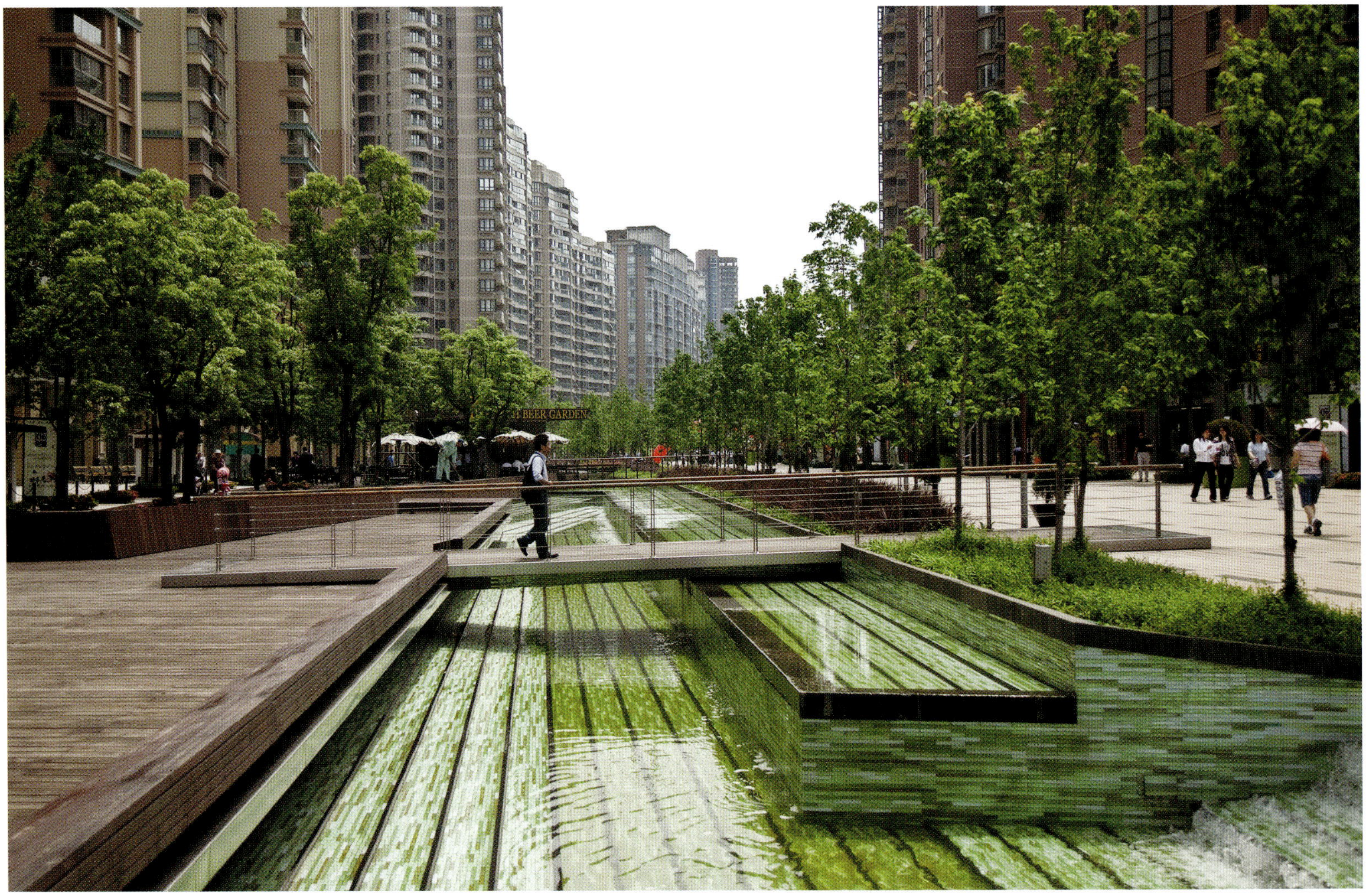

← | Central plaza
↓ | Site plan

↑ | Bustling pedestrian area
↓ | Underground concourse
↘ | Inside-outside landscape continuity

↗ | Roof terraces from above
→ | Public square laid out as ampitheater
↓ | Walk-in fountain lights up at night

D-Cube Landscape City

Seoul

Outside-in design as approach, and inside-out view as esthetic claim, is about arriving and being pulled in: inviting transition for leaving stress behind, landscape penetrating the building, and affordance of outdoor views and feeling indoors creating a comfortable sense of continuity. It is also about going beyond appearance, creating an ordinary visible environment with extraordinary attention to invisible air, sound, memory and feeling. Considering the state of Korea's urban landscape and nature, its cultural traditions, and contemporary ordinary life are important factors for creating people places that are relevant and thus memorable and meaningful, not just functional, but also beautiful and poetic.

PROJECT FACTS

Address: 662, Gyeongin-ro, Guro-ku, Seoul 152-706, South Korea. **Completion:** 2011. **Area:** 28,000 m². **Estimated visitors:** 250,000 per day. **Paving:** imported soft-colored sandstone juxtaposed with Korean granite and pre-cast concrete. **Landscaping:** native Korean trees and wildflowers. **Street furniture:** place-specific landscape and utility structures integrate seating, shelter, and lighting. **Context:** city center shopping area.

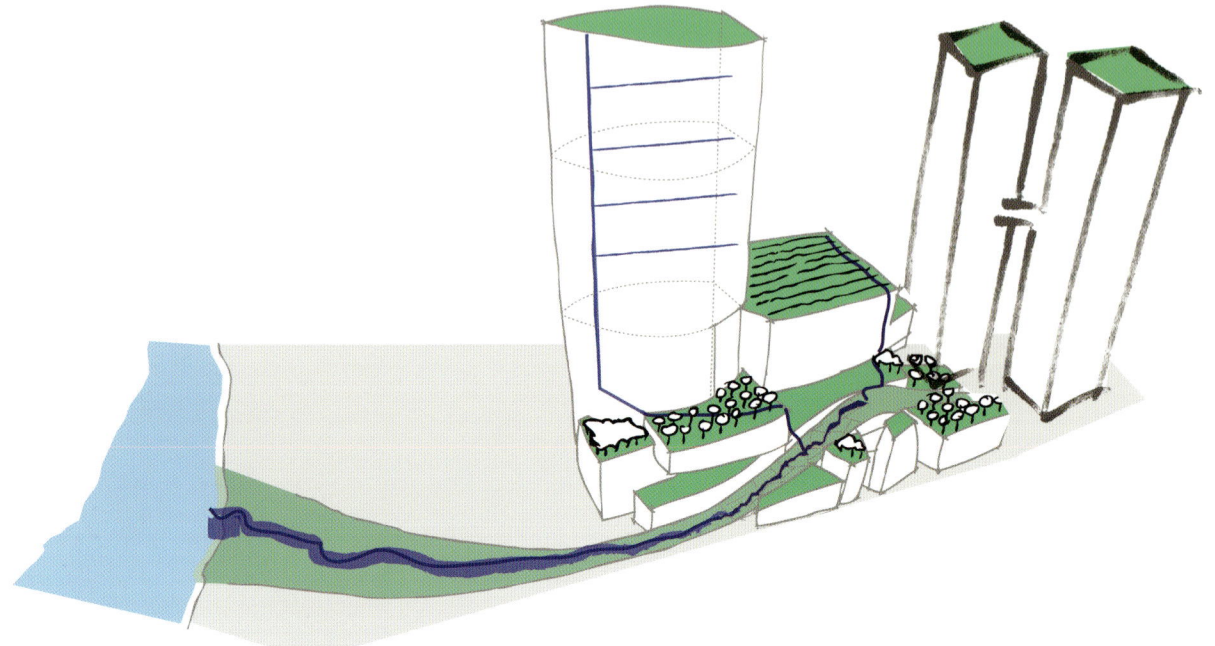

↑ | **Concept sketch,** city as landscape and
building as mountain
↓ | **Building exterior as 'landscape'**

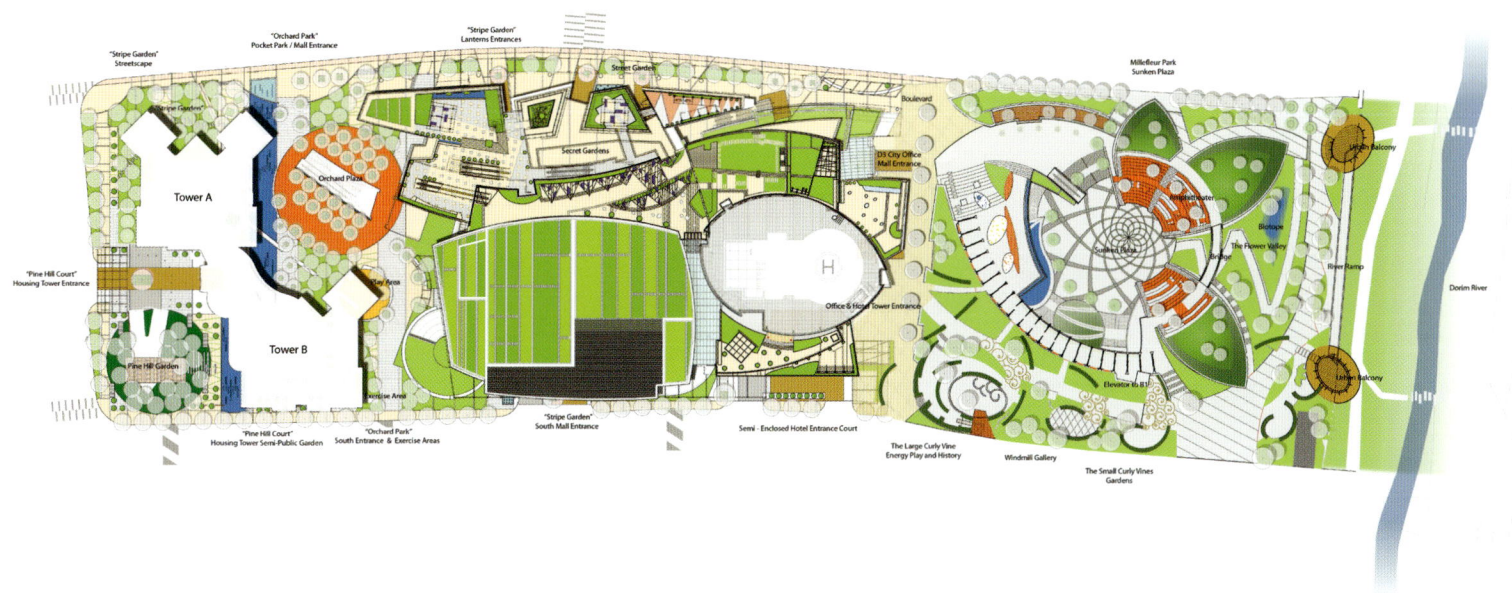

↑ | Site plan
← | Colorful environmental art works give
the design an unique character

James Corner Field
Operations

↑ | **Nicollet Mall Woods and Theater-in-the-round**
↗ | **South Groves,** 'after' image of area
→ | **Nicollet Mall today,** 'before' image of area

Nicollet Mall

Minneapolis

In the early 1900s, Nicollet Avenue was Minneapolis' civic Main Street, with businesses, shop fronts, cars, parking and pedestrian sidewalks. In the 1960s, Lawerence Halprin transformed the street into Nicollet Mall, transit-only thoroughfare, with generous sidewalks and gently curving traffic lanes. Field Operations' new design retains the popular and defining serpentine curve, while creating a greener and more pedestrian-friendly environment. Improvements include grouped groves of trees with movable seating and porous pavement to collect stormwater, eventful lighting, new transit shelters and elegant yet robust paving. A series of integrated features such as a reading room, fire pits and the theater-in-the-round, encourage sociable experiences along the mall and together with the improvements create an elegant urban spine of inter-connectivity, movement, and delight.

PROJECT FACTS

Address: Nicollet Mall, Minneapolis, MN, USA. **Completion:** 2016. **Project team:** Snow Kreilich Architects Inc., Coen + Partners, SRF Consulting Group, Nelson Nygaard, Tillotsen Design Associates, Kestrel Design Group, Cost Construction Services, HTPO. **Area:** 45,000 m². **Context:** commercial pedestrian and transit street.

← | **Loring woods and fire pits**
↓ | **Framework plan,** social seating features

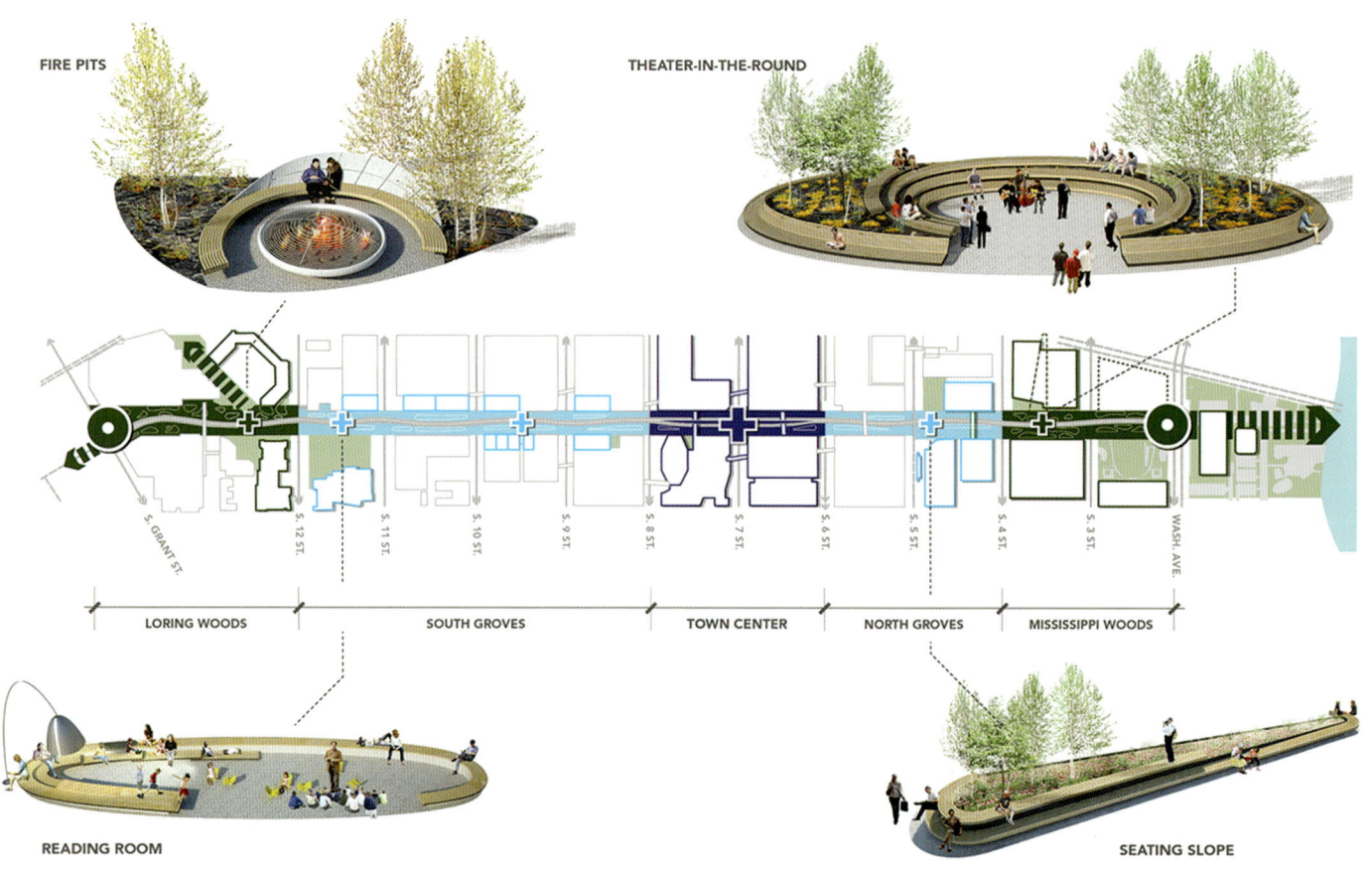

FIRE PITS

THEATER-IN-THE-ROUND

S. GRANT ST.

S. 12 ST.

S. 11 ST.

S. 10 ST.

S. 9 ST.

S. 8 ST.

S. 7 ST.

S. 6 ST.

S. 5 ST.

S. 4 ST.

S. 3 ST.

WASH. AVE.

LORING WOODS

SOUTH GROVES

TOWN CENTER

NORTH GROVES

MISSISSIPPI WOODS

READING ROOM

SEATING SLOPE

↑ | Outdoor café seating in The Groves
← | Paving design

Mikyoung Kim Design

↑ | **Water from above**
↓ | **Canal design,** attraction within the city

↙ | **Illuminated canal**
↓ | **ChonGae Canal**

ChonGae Canal Restoration

Seoul

The ChonGae Canal Restoration Project is an ambitious redevelopment initiative that transformed the urban fabric of Seoul, Korea. This design was the winning project of an international competition and celebrates the source point of cleansed runoff from the city at the start of this seven mile green corridor. The goal was to restore this highly polluted and covered waterway with the demolition of nearly four miles of at grade and elevated highway infrastructure that divided the city. The outcome is the creation of a pedestrian focused zone that brings people to the historic ChonGae River. In addition to the environmental restoration effort, this urban open space has become a central gathering place for the city which is in dire need of more public landscapes.

PROJECT FACTS
Address: 14 Seorin-dong, Jongno-gu, Seoul, South Korea. **Completion:** 2007. **Area:** 91,000 m².
Estimated visitors: 64,000 per day. **Paving:** granite. **Context:** redevelopment initiative in city center.

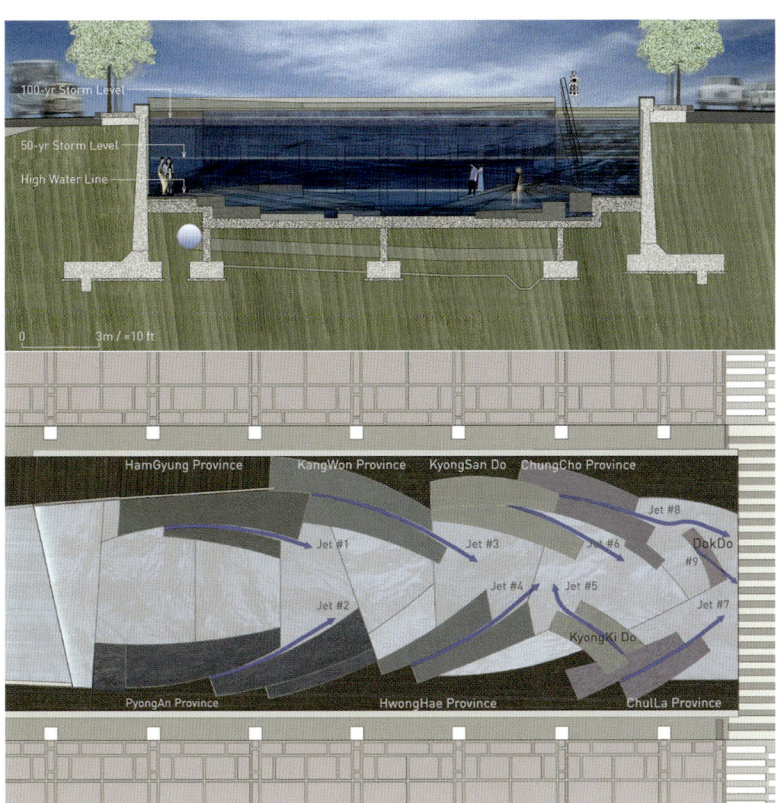

↑ | Section
↓ | Access to the water

↑ | Concrete steps also function as seating
↓ | Children playing in the water

Michael Singer Studio

↑ | Docks at night
↗ | Docks with recreational space
→ | Holes in the dock, mini ecosystems

Living Docks

West Palm Beach

The West Palm Beach Waterfront includes three new docks for boats and a water-taxi to encourage visitors to the downtown area. The central Living Dock is a public promenade with shaded seating areas and innovative in-water planters containing native mangroves, spartina grasses and a visible oyster reef set into the dock. The dock functions as a living system, filtering water and providing small pockets of habitat within an estuarine man-made structure. The new docks were designed to align with the annual Palm Beach Boat Show layout in order to establish permanent circulation spines for the event, reducing the cost and environmental impact of temporary docks for the boat show and other on-water events. Michael Singer Studio led the project design team which included Technomarine, Taylor Engineering, and landscape architect Sanchez and Maddux.

PROJECT FACTS

Address: West Palm Beach Waterfront, FL, USA. **Completion:** 2009. **Landscape architects:** Sanchez and Maddux. **Length:** 130 m. **Area:** 1,200 m². **Landscaping:** red mangroves, spartina grass. **Street furniture:** custom wood benches. **Context:** docks on city waterfront.

← | Oyster bed
↓ | Section

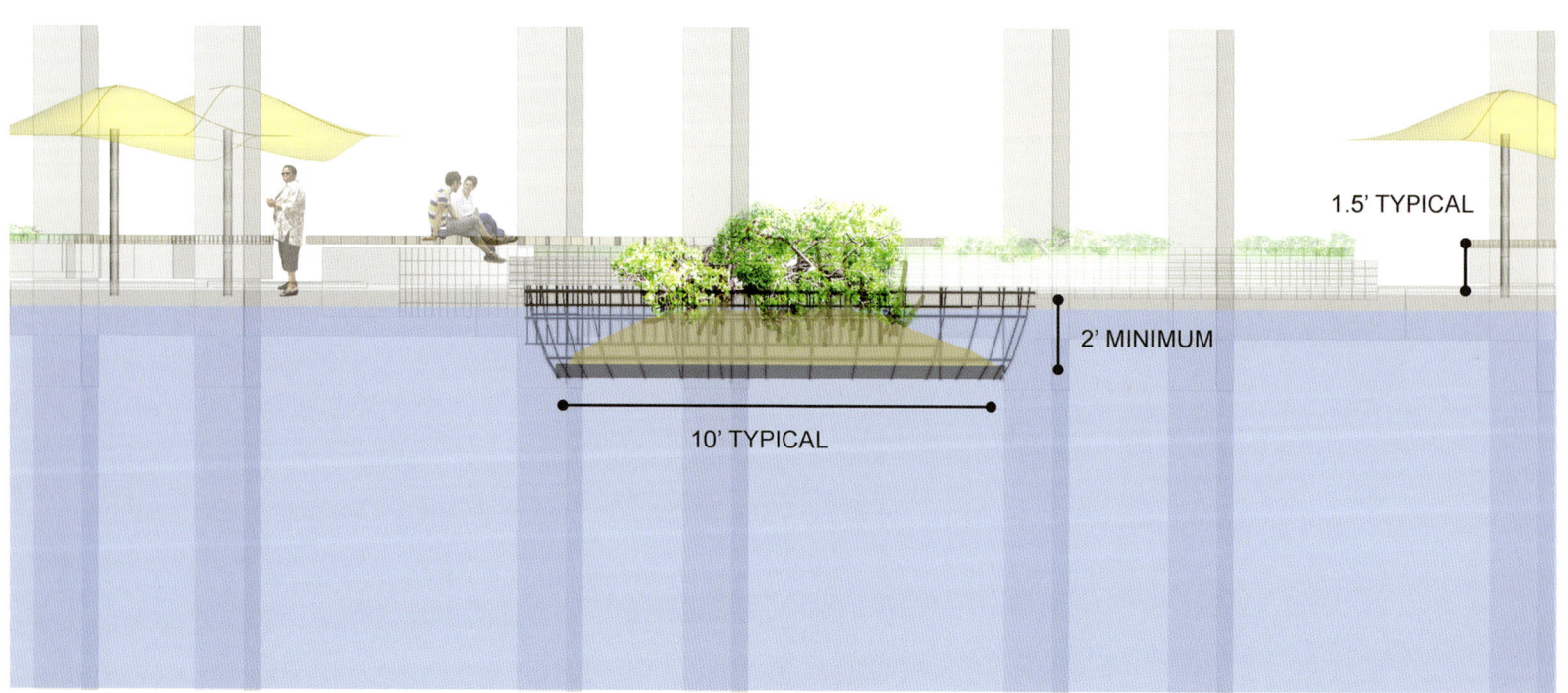

↑ | **Aerial view**
← | **Seating,** integral part of the design

↑ | **Main park**, from above
→ | **Interwoven connections**, boundaries between inside and outside are blurred

Finance Street

Beijing

The Beijing Master Plan provided a historic opportunity to create an international destination in West Beijing. This large project, which includes a mix of uses – housing, retail, hotel, office and cultural facilities, is focused on a central park known as "The Heart" of Western Beijing. Approximately 90 percent of the project area was located over garage structure and required a special approach in thinking about planting and landscape construction. SWA's work paralleled that of SOM, the building architect, and integrated urban design and landscape architecture into both the physical and cultural structure of the design. A large civic plaza, which features a computer-animated fountain and light show, fronts Finance Street while a series of intimate courts center each urban block.

PROJECT FACTS

Address: Finance Street, Central Park District, Beijing, China. **Completion:** 2008. **Architects and engineers:** Skidmore, Owings & Merrill LLP. **Area:** 404,686 m². **Landscaping:** plant palette is restricted to indigenous materials. **Context:** urban pedestrian zone.

← | **Pedestrian streetscape**, retail and cafés
↓ | **Central fountain**

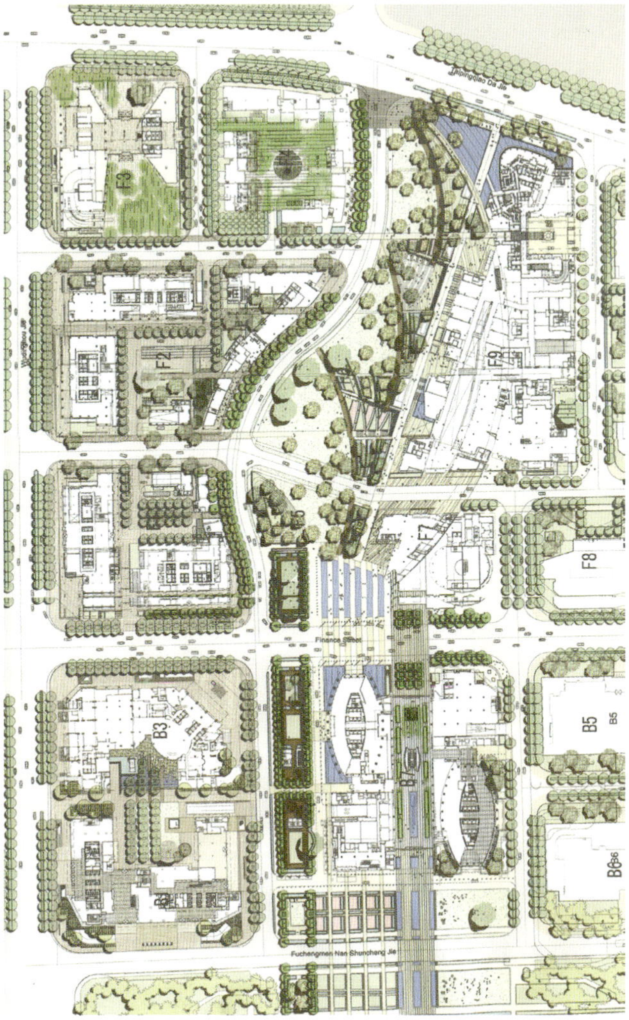

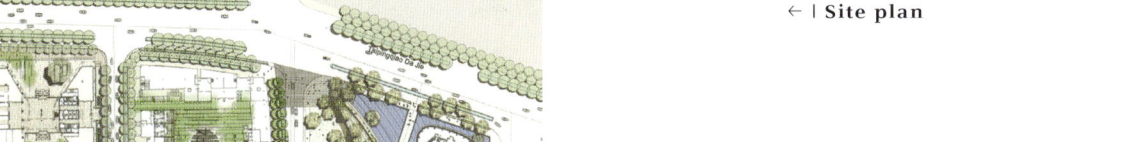

↑ | Pools at the base of the hotel
← | Site plan

↑ | **National plaza**, waterfront gateway
↗ | **National Harbor central space**
→ | **Gateway to the Potomac River**

National Harbor

Washington DC

National Harbor is a mixed-use community on the Potomac River just south of Washington DC. While conveniently located near key tourist sites, National Harbor is a resort and convention destination that offers an alternative to the urban experience of Washington proper. Sasaki provided urban planning and landscape architecture for principal exterior spaces of the community, architecture for key buildings, and graphic design for signage and wayfinding systems.

PROJECT FACTS

Address: 165 Waterfront Street, National Harbor, Washington DC, Md 20745, United States. **Landscape architects:** LandDesign Inc., TWS Design, Inc. **Completion:** 2008. **Area:** 603,870 m². **Estimated visitors:** 20,000 per day. **Paving:** cobblestones, concrete unit pavers, granite. **Context:** city waterfront.

← | Waterfront terrace
↓ | Site plan

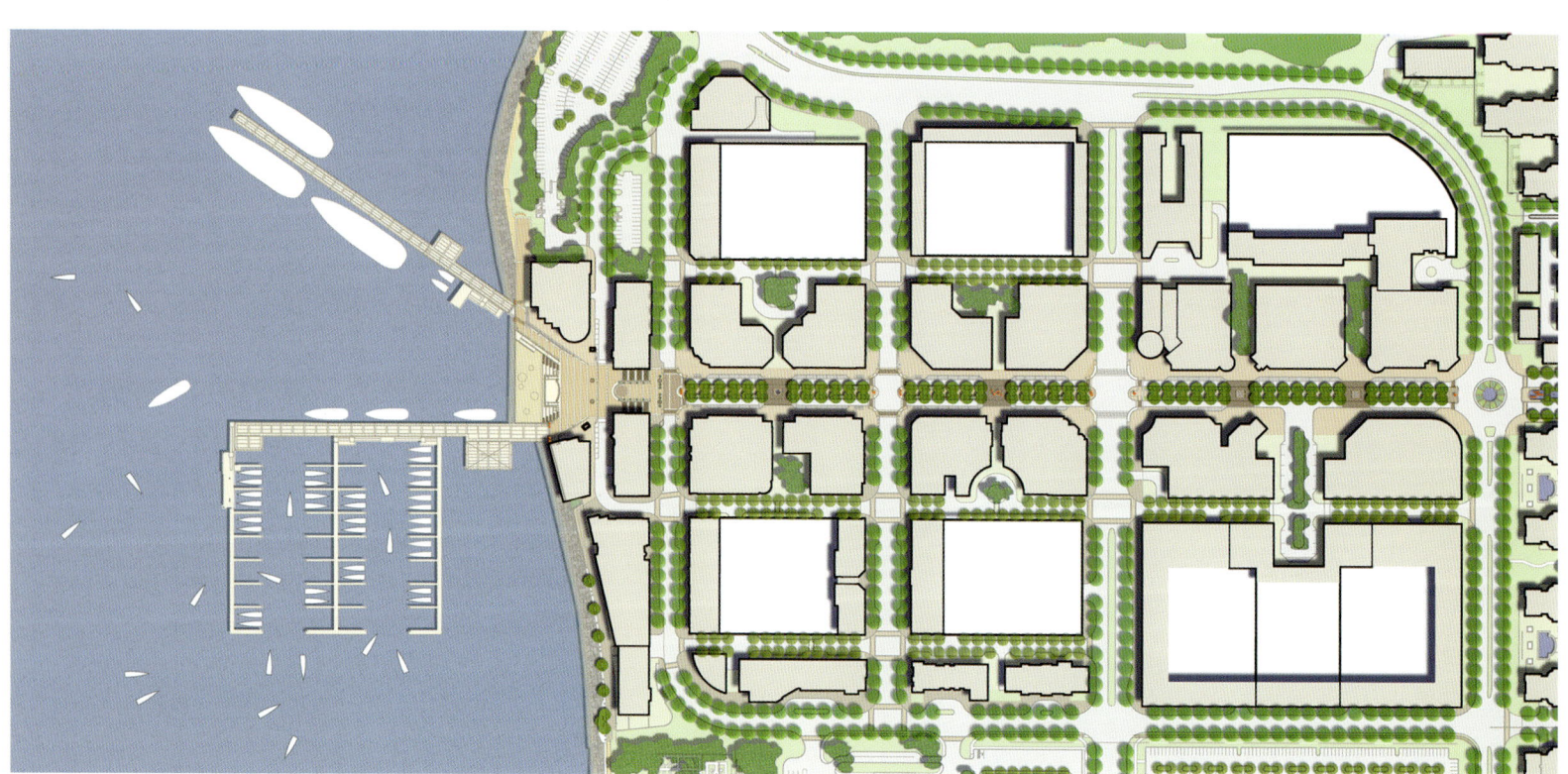

↑ | **Stairs**, Potomac River
← | **Seating area**

↑ | **Aerial view**
→ | **Playground**

Hauser Plads

Copenhagen

Hauser Plads is a square in the center of Copenhagen, Denmark. It is part of the public space around the Købmagergade shopping street that was designed by Karres en Brands. The square features a recreational landscape formed by green 'recreation hills'. Facilities will include showers, changing rooms, a canteen and parking spaces. Moreover, offices and a meeting area are also housed here, these have to provide a pleasant working environment for their users. Construction of a patio at cellar level allows natural light to enter and creates an outdoor area for the employees. The patio's curved glass walls maximize the views from the indoor working area. Various types of planting change their color and texture throughout the seasons, and the flagstone paving area creates places for rest breaks whilst serving as an outdoor workplace.

Address: Hauser Plads, Copenhagen, Denmark. **Completion:** 2013. **Area:** 3,420 m². **Landscaping:** integrated play and green areas. **Context:** public square and play area in city center.

← | **Interior,** surrounded by pedestrian area
↓ | **Site map**

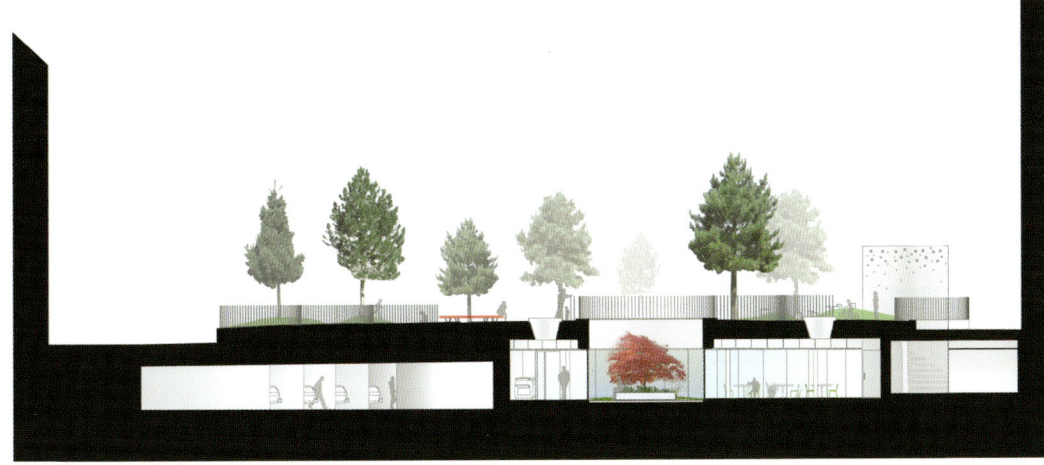

↖ | Section
↙ | Patio of the "Center for Renhold"

↑ | **Galleria area**, creek
→ | **City Creek Center**, bird's-eye view

City Creek Center

Salt Lake City

SWA provided landscape architecture and urban design services for City Creek Center, formerly the site of two single-use enclosed malls City Creek Center is a redevelopment employing a bold landscape strategy to anchor 536 residences, offices and retail space. Bisecting three of Salt Lake City's 201-meter superblocks, the new pedestrian realm is built over structure and inserts a novel urban design strategy at a new scale into the existing city fabric. Hundreds of native trout thrive in the authentic recreation of the historic City Creek, which is the Center's iconic feature. Using more than 50 percent of the debris recycled from previous structures, City Creek Center is the first shopping center to achieve LEED Silver pilot program certification. All four residential towers associated with the development garnered LEED Gold status.

PROJECT FACTS

Address: 50 South Main Street, Salt Lake City, UT 84144, USA. **Completion:** 2012. **Retail architects:** Callison Architects in partnership with Hobbs and Black. **Area:** 101,171 m². **Landscaping:** native trees and plants. **Context:** residential, office and retail zones.

↖ | Galleria area
↓ | Tranquil space to relax

↑ | Oval court at night
← | Creek winds through restaurants and shopping area

Index

Archit

A

Architects' Index

3:0 Landschaftsarchitektur Gachowetz Luger Zimmermann

Nestroyplatz 1/1
1020 Vienna (Austria)
T +43.1.9690662
office@3zu0.com
www.3zu0.com

→ **158**

José Adrião Arquitectos

Rua Gilberto Rola 41
1350-154 Lisbon (Portugal)
T +351.213.620762
ja@joseadriao.com
www.joseadriao.com

→ **20**

Aspect Studios
Sacha Coles, Scott van den Boogaard, Saskia van der
Put, Catherine Wilson, Kate Luckraft, Thea Harris,
Emma Cave, Liew Kheng Teik, Joel Munns

Studio 61, Level 6, 61 Marlborough Street
Surry Hills, 2010 (Australia)
T +61.2.96997182
F +61.2.96997192
sydney@aspect.net.au
www. aspect.net.au

→ **108, 172, 200**

AWP office for territorial reconfiguration
Matthias Armengaud, Alessandra Cianchetta, Marc
Armengaud

25, rue Henry Monnier
75009 Paris (France)
T +33.1.53209215
awp@awp.fr
www.awp.fr

→ **162**

Behnisch Architekten
Stefan Behnisch, Robert Hösle, Stefan Rappold, Robert
Matthew Noblett

Rotebühlstraße 163A
70197 Stuttgart (Germany)
T +49.711.607720
F +49.711.6077299
ba@behnisch.com
www.behnisch.com

→ **28**

BHF Bendfeldt Herrmann Franke Landschaftsarchitekten
Jens Bendfeldt, Uwe Herrmann, Uli Franke

Jungfernstieg 44
24116 Kiel (Germany)
T +49.431.997960
bendfeldt@bhf-ki.de
www.bhf-ki.de

→ **44**

BIG

Kløverbladsgade 56
2500 Copenhagen (Denmark)
T +45.7221.7227
F +45.3512.7227
big@big.dk
www.big.dk

→ **114, 124**

birke . zimmermann landschaftsarchitekten
Florian Birke, Claudia Zimmermann

Wichertstraße 5
10439 Berlin (Germany)
T +49.30.48494019
info@birkezimmermann.de
www.birkezimmermann.de

→ **32**

BPG Landschaftsarchitekten
Norbert Kerl, Andrea Ziegenrücker, Jochen Kehm

Karlstraße 20
35444 Biebertal (Germany)
T +49.6409.81070
F +49.6409.810730
info@bpg-biebertal.de
www.bpg-biebertal.de

→ **40**

Claude Cormier et Associés

1223, rue des Carrières, studio A
Montréal, Québec (Canada)
T +1.514.8498262
F +1.514.2798040
info@claudecormier.com
www.claudecormier.com

→ 122

James Corner Field Operations

475 10th Avenue, 9th floor
New York City, NY 10018 (USA)
T +1.212.4331450
www.fieldoperations.net

→ 212

Design Workshop
Steven Spears, Philip Koske, Alex Ramirez, Kelan Smith

800 Brazos Street, Suite 490
Austin, TX 78701 (USA)
T +1.512.4990228
dwi@designworkshop.com
www.designworkshop.com

→ 104

Lola Domènech

Ronda Sant Pere 58 3 2
08010 Barcelona (Spain)
T +34.932.683277
F +34.932.683277
ld@loladomenech.com
www.loladomenech.com

→ 62

Earthscape
Eiki Danzuka

5-15-15 Kamiuma Setagaya
154-0011 Tokyo (Japan)
T +81.3.64509588
F +81.3.64509544
info@earthscape.co.jp
www.earthscape.co.jp

→ 154

Entasis Arkitekter

Flæsketorvet 75
1711 Copenhagen (Denmark)
T +45.33.339525
entasis@entasis.dk
www. entasis.dk

→ 134

Gillespies

1 St John's Square
London EC1M 4DH (England)
T +44.20.72532929
www.gillespies.co.uk

→ 36

Gustafson Guthrie Nichol

1932 First Avenue, Suite 700
Seattle, WA 98101 (USA)
T +1.206.9036802
contact@ggnltd.com
www.ggnltd.com

→ 88, 166

Hassell
Angus Bruce

Level 2, Pier 8/9, 23 Hickson Road
2000 Sydney (Australia)
T +61.2.91012000
sydney@hassellstudio.com
www.hassellstudio.com

→ 128

hirner & riehl architekten und stadtplaner bda
Martin Hirner, Martin Riehl

Holzstraße 7
80469 Munich (Germany)
T +49.89.218984430
F +49.89.2189844333
info@hirnerundriehl.de
www.hirnerundriehl.de

→ 76

In Situ Atelier de paysage et urbanisme

8, quai Saint Vincent
69001 Lyon (France)
www.in-situ.fr

→ 54

Isthmus Group
David Irwin, Tim Fitzpatrick, Nada Stanish

43 Sale Street, Freemans Bay
Auckland (New Zealand)
akl@isthmus.co.nz
www.isthmus.co.nz

→ 48

Karres en Brands landscape architecture + urban planning

Oude Amersfoortseweg 123
1212 AA Hilversum (The Netherlands)
T +31.35.6422962
info@karresenbrands.nl
www.karresenbrands.nl

→ 180

KBP.EU (Karres en Brands + Polyform)

Oude Amersfoortseweg 123
1212 AA Hilversum (The Netherlands)
T +31.35.6422962
info@karresenbrands.nl
www.karresenbrands.nl

→ 16, 230

Keller Damm Roser
Landschaftsarchitekten Stadtplaner
Regine Keller, Mattias Roser, Franz Damm

Dachauer Straße 17
80335 Munich (Germany)
T +49.894.423170
info@keller-damm-roser.de
www.keller-damm-roser.de

→ 24

Mikyoung Kim Design

119 Braintree Street Suite 103
Boston MA 02134 (USA)
T +1.617.7829130
F +1.617.7826504
office@myk-d.com
www.myk-d.com

→ 216

Knippers Helbig

Tübinger Straße 12–16
70178 Stuttgart (Germany)
T +49.711.24839360
F +49.711.2483936 88
stuttgart@knippershelbig.com
www.knippershelbig.com

→ 118

Hans-Hermann Krafft

Überseestraße 27
10318 Berlin (Germany)
T +49.30.6141303
hh.krafft@wes-la.de

→ 50

L35 Architects
Jos Galán, Eduardo Simarro, Caterina Memeo

Avinguda Diagonal 466 6°
08006 Barcelona (Spain)
T +34.93.2922299
F +34.93.4160530
L35@L35.com
www.L35.com

→ 188

Landscape Projects

31 Blackfriars Road
Salford M37AQ (England)
T +44.161.839.8336
F +44.161.8397155
post@landscapeprojects.co.uk
www.landscapeprojects.co.uk

→ 66

Martínez Lapeña Torres Arquitectos
José Antonio Martínez Lapeña, Elías Torres

Carrer Roca i Batlle 14
08004 Barcelona (Spain)
T +34.93.2121416
F +34.93.2540682
jamlet@coac.net
www.jamlet.net

→ 56

M&N Environmental Planning Institute
Koshi Ohashi

2-3-21 Soshigaya Setagaya-ku
Tokyo 157-0072 (Japan)
T +81.3.34836165
F +81.3.34831767
m-and-n@mn-epi.co.jp
www.mn-epi.co.jp

→ 198

Mariñas Arquitectos Asociados
José C Mariñas

PO Box 66
41450 Seville (Spain)
T +34.615.350284
torrebabel@torrebabel.com
www.torrebabel.com

→ 150

Mobimo

7, rue de Genève
1003 Lausanne (Switzerland)
T +41.21.341121215
F +41.21.3411213
communication@mobimo.ch
www.mobimo.ch

→ 12

Moore Ruble Yudell Architects and Planners
Buzz Yudell, Michael Martin, Neal Matsuno

933 Pico Boulevard
Santa Monica, CA 90405 (USA)
T +1.310.4501400
F +1.310.4501403
info@mryarchitects.com
www.moorerubleyudell.com

→ 136

Novell Tullett
Simon Lindsley ,Ian Richardson, Isabelle Carter, Oliver Bence, Katherin Westphal, Jane Fowles, Jackie Shore, Anwen Victory, Carol Haines, Ben Oakman, Jochen Rabe, Thomas Schneider

The Old Mess Room, Barrow Court Lane
Bristol BS48 3RW (United Kingdom)
T +44.1274.462476
bristol@novelltullett.co.uk
www.novelltullett.co.uk

→ 22

OikosDesign
Anemone Beck Koh

Bosrandweg 5
6703 EB Wageningen (The Netherlands)
T +31.317.410508
F +31.317.450698
oikos@oikosdesign.nl
www.oikosdesign.nl

→ 208

OKRA landschapsarchitecten
Martin Knuijt, Wim Voogt, Boudewijn Almekinders

Oudegracht 23
3511 AB Utrecht (The Netherlands)
T +31.30.2734249
F +31.30.2735128
mail@okra.nl
www.okra.nl

→ 146

Atelier Jacqueline Osty & Associés

77, rue de Charonne
75011 Paris (France)
T +33.14.3486384
atelier@osty.fr
www.osty.fr

→ 184, 192

r+b landschafts architektur
Sonja Rossa-Banthien, Jens Rossa

Königstraße 12
01097 Dresden (Germany)
T +49.351.8107505
F +49.351.8107504
buero@rplusb.de
www.rplusb.de

→ 60

Sasaki Associates

64 Pleasant Street
Watertown, MA 02472 (USA)
T +1.617.926.3300
lmen@sasaki.com
www.sasaki.com

→ 140, 236

SBA International

Leuschnerstraße 25
70176 Stuttgart (Germany)
T +49.711.25859530
info@sba-int.com
www.sba-int.com

→ 118

scape Landschaftsarchitekten
Rainer Sachse, Hiltrud Lintel, Matthias Funk

Friedrichstraße 115a
40217 Düsseldorf (Germany)
T +49.211.3020370
F +49.211.30203720
post@scape-net.de
www.scape-net.de

→ 84

Schønherr
Rikke Juul Gram, Nina Jensen, Torben Schønherr

Klosterport 4A
8000 Aarhus (Denmark)
T +45.86.186900
mail@schonherr.dk
www.schonherr.dk

→ 160

Michael Singer Studio
Michael Singer, Jason Bregman, Jonathan Fogelson

321 North West 1st Avenue
Delray Beach, Fl 33444 (USA)
T +1.561.8657683
info@michaelsinger.com
www.michaelsinger.com

→ 176, 218

AS&P - Albert Speer & Partner
Friedbert Greif, Prof. Albert Speer, Johannes Dell, Axel Bienhaus, Michael Denkel, Michael Dinter, Gerhard Brand, Joachim Schares, Stefan Kornmann

Hedderichstraße 108–110
60596 Frankfurt/Main (Germany)
T +49.69.6050110
F +49.69.605011500
mail@as-p.de
www.as-p.de

→ 132

ST raum a.
Stefan Jäckel, Bernd Kusserow, Uwe Großkopf, Tobias Micke, Oliver Alten, Katrin Klingberg

Waldemarstraße 33 A
10999 Berlin (Germany)
T +49.30.6166090
F +49.30.61660917
info@strauma.com
www.strauma.com

→ 100, 92

SWA Group
René Bihan, Bill Callaway, Gerdo Aquino, Ying-Yu Hung

2200 Bridgeway Boulevard
Sausalito, CA 94965 (USA)
T +1.415.3325100
sausalito@swagroup.com
www.swagroup.com

→ 204, 222, 234

Tabanlıoğlu Architects
Murat Tabanlıoğlu, Melkan Gürsel Tabanlıoğlu

Mesrutiyet Caddesi 67
34430 Istanbul (Turkey)
T +90.212.2512111
info@tabanlioglu.com
www.tabanlioglu.com

→ 196

TCL

385 Drummond Street
Melbourne (Australia)
T +61.393.804344
melb@tcl.net.au
www.tcl.net.au

→ 70

Tonkin Liu

5 Wilmington Square
London WC1X 0ES (England)
T +44.20.78376255
mail@tonkinliu.co.uk
www.tonkinliu.co.uk

→ 144

Topotck 1
Lorenz Dexler, Martin Rein-Cano

Sophienstraße 18
10178 Berlin (Germany)
T +49.30.2462580
F +49.30.24625899
topotek1@topotek1.de
www. topotek1.de

→ 114

Valentien + Valentien
Christoph Valentien, Donata Valentien

Hauptstraße 42
82234 Weßling (Germany)
T +49.8153.952010
F +49.8153.952014
valentien@valentien.de
www.valentien.de

→ 96

Jörn Wagner
Freier Landschaftsarchitekt
Jörn Wagner

Holtenauer Straße 94
24105 Kiel (Germany)
T +49.431.35052
F +49.431.35053
landschaftsarchitekt.wagner@t-online.de
www.wagner-la.de

→ 74

WES LandschaftsArchitektur
Peter Schatz, Henrike Wehberg-Krafft, Hinnerk
Wehberg, Michael Kaschke, Wolfgang Betz

Reichenberger Straße 124
10999 Berlin (Germany)
T +49.30.58584440
berlin@wes-la.de
wes-la.de

→ 50, 80, 166

PICTURE CREDITS

Fernando Alda/www.fernandoalda.com 150–153
Emanuel Andel 158 (portrait)
Shigeki Asanuma 156
Iwan Baan 114, 115, 116 b.
Peter Bennetts, Melbourne 128–131
BESCO GmbH 92 a. 95 a. r., 100 b. l., b. r., 101, a.
BHF Landscape Architects 44–47
© Maciej Bledowski – Fotolia.com 112–113
Ute Boeters 74 (portrait)
Alexander Paul Brandes 50, 51, 53
Jens Braune del Angel 132 (portrait)
Marcus Bredt 92 b. 93, 94 a. l. b., 100 a., 101 b., 102, 103
Katja Brügmann 75 a. l., a. r.
Catalfumo Construction 178
Oscar Chinarro 188 a., 191
Claude Cieutat, Paris 193 b., 194, 195
City of West Palm Beach 221 a.
Gregori Civera 162 (portrait)
Keith Collie 144, 145 a. r., b.
Marc Cramer 122, 123, a., b., r.
Simon Devitt 48, 49
Doug and Wolf 200–203
Guy Durand 184 (portrait), 192 (portrait)
C. Ernst, Dresden 60 (portrait)
Torben Eskerod 116 a.
FG+SG 20–23
Tom Fox, SWA Group 204–207, 234
Dubois Fresney, Paris 192, 193 a.
Jiro Fukasawa 154 (portrait)
Gillespies/John Cooper 36–39
Adrià Goula 62–65
Florian Groehn 172–175
© grondetphoto – Fotolia.com 170–171
Andreas Harbach, Lifestylephoto 79 b.
Michael Heinrich, Munich 24
Carsten Ingemann 160 a., b., r., 161 a.
Jamlet, Barcelona 56 (portrait)
Lourdes Jansana 56–59
Ulrik Jantzen 124 (portrait)
Vincent Jendly 12, 14 b.
Hanns Joosten 114 (portrait)
Tine Juel 134 (portrait)
Gisbert Jungermann 75 b.
Karres en Brands 180–182
John Keatley 88 (portrait), 164 (portrait)
Milo Keller 13 b.
Taeoh Kim 216 b., 217 a. r.
Jusuck Koh 208 (portrait)
Jörg Koopmann, Munich 24 (portrait), 25, 26, 27
Goran Kosanovich 204 (portrait l.)
Craig Kuhner, Arlington 226, 227 b., 228, 229 b., 235, 236, 237
© Tiago Ladeira – Fotolia.com 10–11
Michael Latz 96–99
Will Lew 122 (portrait)
Colins Lozada 136, 137, 139
Mike Magnussen 117 b.
David Matthiessen 28, 30
Thomas Mayer 196–199
Arnaud Meylan 15
Michael Nicholson 70–73

L. Michow & Sohn GmbH, Hamburg 61 b.
Helge Mundt 52
Neoscape, New York 91
Frank Ockert 29, 31
Elise Oddy 128 (portrait)
Koji Okumura/Forward Stroke 154 a., 155, 157 b.
Marc Ostow 28 (portrait)
Thomas Ott, Mühltal/www.o2t.de 118–121
Guillaume Paradis 123 b. l.
Kieith Parry 22–25
M. Pavlitzek, Lingen 60 a.
Pascal Pernix 138
Nicolás Pinzón 188 b., 189, 190
Anna Posilano 163 a.
Frank Richter 92 (portrait), 100 (portrait)
Axel Schön 44 (portrait)
Shaw + Shaw, London 66–69
Ty Stange 16–19 b., 230–233
David Stansbury 176, 177, 179, 218 (portrait), 219, 221 b.
Robert Such 216 a., 217, b.
Bill Tatham 222 (portrait), (portrait l.)
Ingolf Timpner 50 (portrait), 80 (portait), 166 (portrait)
Chris van Uffelen 6, 8, 9 b.
André Wagner 13 a.
Rolf Wegst, Gießen 40–43
Wikifrits / Wikipedia Commons 9 a.
Ed Wonsek 227 a., 229 a.
Simon Wood 108–111
Bruno Simões, Lisbon 21 (portrait)

All other pictures, especially portraits and plans, were made available by the architects.

Cover front: Iwan Baan
Cover back: left: Vincent Jendly
right: Simon Wood

IMPRINT

The Deutsche Nationalbibliothek lists this publication in the Deutsche Nationalbibliografie; detailed bibliographic data are available in the Internet at http://dnb.dnb.de.

ISBN 978-3-03768-190-9

© 2015 by Braun Publishing AG
www.braun-publishing.ch

1st edition 2015

Project coordination: Editorial Office van Uffelen
Editorial staff: Carina Kaminski, Constanze Lauer, Lisa Rogers
Graphic concept: ON Grafik | Tom Wibberenz
Layout: Carina Kaminski, Constanze Lauer, Lisa Rogers